WEATHER PREDICTION
WHAT EVERYONE NEEDS TO KNOW®

WEATHER PREDICTION

WHAT EVERYONE NEEDS TO KNOW®

ROBERTO BUIZZA

OXFORD
UNIVERSITY PRESS

Oxford University Press is a department of the University of Oxford. It furthers
the University's objective of excellence in research, scholarship, and education
by publishing worldwide. Oxford is a registered trade mark of Oxford University
Press in the UK and certain other countries.

"What Everyone Needs to Know" is a registered trademark of
Oxford University Press.

Published in the United States of America by Oxford University Press
198 Madison Avenue, New York, NY 10016, United States of America.

CIP data is on file at the Library of Congress

ISBN 978–0–19–765212–1 (pbk.)
ISBN 978–0–19–765213–8 (hbk.)

DOI: 10.1093/wentk/9780197652138.001.0001

Paperback printed by Sheridan Books, Inc., United States of America
Hardback printed by Bridgeport National Bindery, Inc., United States of America

To Gioia, Caterina, Leonardo, and Ludovico

CONTENTS

3 Observing the Earth system   44

4 Modeling the Earth system   72

PREFACE

Weather has always affected human life. Understanding how weather events form and predicting what kind of weather is coming can help enormously to manage weather risk, and it will become even more important as we shift toward more strongly weather-dependent energy sources.

Some big steps forward in numerical weather prediction have been made in the past 40 years, thanks to advances in four key areas: the way we observe the Earth, the scientific understanding of the phenomena, high-performance computing (that has allowed the use of increasingly complex models), and modeling techniques.

Thanks to advances in these four areas, in the last forty years we have improved short- and medium-range forecasts by about 1 day per decade: forecasts issued today and valid for 5 days from now, are as accurate as forecasts issued forty years ago and valid for tomorrow. Forty years ago we were not able to issue accurate forecasts beyond a few days, while today we issue accurate monthly and seasonal forecasts. Today we are capable of predicting extreme events such as hurricanes and extra-tropical windstorms very accurately up to 7–10 days ahead. We can predict the most likely path and intensity of storms before they hit a community, estimate the confidence level of the forecast, and give very valuable indications of their probable impact. Larger-scale phenomena that affect entire countries, such as heat or cold waves, periods with extremely high or low temperatures lasting for days, can be forecast up to 2–3 weeks before the events occur. Phenomena that affect a big portion of the oceans or of a continent and that evolve slowly, such as the warming of the sea-surface temperature in the Pacific Ocean when an El Niño event occurs, can be predicted months ahead, and in some cases even further in advance.

In the future, weather prediction will play an ever more important role, due to the impact of the human-caused climate change and the transformations of the human activities we are going to implement to address it. A warmer, climate-change-induced future sees an increasing frequency of extreme events, and thus the ability of being able to predict them even earlier than today is extremely important to manage weather risk. In the future, electricity and energy production is going to rely more than today on weather-sensitive sources like solar, wind, and hydropower stations, and thus more accurate weather forecasts will allow a better planning and matching of energy supply and demand.

This book discusses some of the key topics linked to weather prediction. Like the other books in the What Everyone Needs to Know™ series, it is intended for the general public, including policymakers, environmentalists, students, and scientists in any field. It has been written for those who know a little about weather, the physics of the atmosphere or the ocean, satellites, or supercomputing. I expect that the interested reader should be able to follow this fascinating journey into numerical weather prediction, an activity that has been helping humans to live better. After reading this book, the curious reader who wants to learn more about the subjects that I discuss could consult the Further Reading list, and the articles published in the peer-reviewed literature dealing with weather and climate science.

I have tried my best to answer some key questions that I and fellow colleagues have been asked over the years: How are weather forecasts generated? How complex are the models used in numerical weather prediction, and how do we solve them? Was this event predictable? Why was this forecast wrong? How did we manage to predict this hurricane path 10 days before the event? Will weather forecast continue to improve, or is there a predictability limit?

The book starts with an introduction to weather and climate, and a description of the spatial and temporal characteristics of weather phenomena. We will discuss the role of observations in understanding weather events, and their characteristics, accuracy, and uncertainty are described. We will then illustrate how we derive the equations that can be used to construct a model of the Earth system. Chapters 5, 6, and 7 discuss numerical weather prediction, how forecasts are generated, the sources of model errors, and why uncertainties will always affect weather forecasts. We will describe the ensemble methods that we have developed to estimate the forecast uncertainties and to generate probabilistic forecasts, methods that are today used also to generate climate projections. Chapter 8 introduces the concept of the forecast skill horizon.

Since climate change and climate science developments have been affecting weather prediction, we will also discuss the links between these two areas of science, and how weather and climate scientists have been working closely together to advance our understating of the Earth system and improve our prediction methods. Finally, we will illustrate how numerical weather prediction can be further improved, the forecast skill horizon extended, and new types of products designed.

The approach that I followed in all these chapters is to be physically correct and to use the right concepts and logical arguments to support my statements. In some parts, I have introduced and discussed the meaning of a few, key equations to show the reader how they are written and to explain how they are solved, although in general I have avoided discussing them.

Each chapter starts with a list of the questions it addresses and ends with a summary of the key points that it discussed. At the end of the book, I have inserted a list of recent books to which a reader interested in learning more about the science behind weather prediction and climate change can refer. I have also listed a few, key websites of leading institutions working in numerical weather prediction and climate that I have mentioned in the text.

Let me tell you a bit about myself. I have enjoyed immensely working in numerical weather prediction. Understanding weather phenomena, modeling them, and trying to predict their evolution have been part of my life for over 35 years. I was extremely lucky to end up working in this field at the European Centre for Medium-Range Weather Forecasts (ECMWF), the worldwide leader in global, medium-range weather prediction. There I learned that complex problems can be addressed with effective teamwork that brings together experts from many different disciplines, all determined to share their energy and thoughts to advance the scientific understanding and solve the challenges that are encountered every day.

In this field, problems to solve are difficult and challenging, and progress is slower than you would wish, but the impact that one can make by advancing scientific understanding is huge. One of the best aspects of this work has been that any advance we made, any little improvement in forecast skill that we achieved via our research and development, had an immediate and tangible impact on society.

There are still many open questions that could, if answered, help provide better forecasts, especially in the medium range (forecasts valid 5–15 days ahead), and the monthly and seasonal time scales. On all these forecast ranges we can still make progress, saving lives and reducing damages linked to extreme events—extremes that climate change has been making more frequent and intense than in the past.

I encourage young and curious minds to think about working in weather and climate science: you do not need to be a meteorologist or an expert in the physics of the atmosphere to contribute and have an impact. Experts in many different fields have and will continue to contribute to advances in these fields.

I hope you will enjoy the reading.

ACKNOWLEDGMENTS

This book is based on what I learned while working at the European Centre for Medium-Range Weather Forecasts (ECMWF) between 1991 and 2018. I would like to acknowledge the ECMWF family, first for deciding in 1991 to take me on board, and then for giving me the honor and privilege to remain in the family for many years. Working with you all has been fun, challenging, and at times stressful, but it has surely given me a unique opportunity to contribute to improving weather prediction. I also want to thank the many friends and colleagues at national weather services, universities, and research institutes on all continents, with whom I have been interacting since 1991. The way we have been openly exchanging ideas and comparing notes, and the continuous challenges we have been putting to each other, are the key reasons why weather prediction has been improving at a very impressive rate.

In this book, I have done my best to answer a wide range of questions on weather predictions that touch on different areas of research and operational production. Many of these questions deal with topics on which I have worked with you all in the past. Without our interactions, and the many projects we realized together, I would not have been able to answer them.

Let me thank once more ECMWF for having allowed me to use many of their figures in this book.

I am also grateful for the editorial team at Oxford University Press for making the process of publishing this book smooth and enjoyable. Special thanks to Koperundevi Pugazhenthi, Jeremy Lewis, Lilith Dorko, and Michelle Kelley at OUP. I am also grateful to the reviewers for their comments and feedback on my early proposal.

Last, and most importantly, let me thank my wife and (grownup) children for their fantastic support and the challenging discussions we had, everywhere and at the most diverse times of the days and nights, on so many issues in mathematics, physics, and life!

1

WEATHER AND CLIMATE

In this chapter we define the meaning of the two words *weather* and *climate*, how weather phenomena determine the climate, the role that phenomena with different spatial and temporal scales play, and the key weather variables. More specifically, we will be addressing the following questions:

1. What is the key difference between weather and climate?
2. Do weather and climate vary spatially and temporally?
3. Is there a clear separation between weather and climate?
4. Is weather affected by all (fast and small-scale and slow and large-scale) phenomena?
5. Which coordinate system is used to study atmospheric and oceanic motions?
6. What are the key weather variables?
7. Why does weather change?
8. How are motions in the atmosphere generated?

1.1 What is the key difference between weather and climate?

Weather is what we experience every day, while climate describes the statistics of weather phenomena computed over a long period of time, say at least a season, ideally a few years.

With the word *statistics* we mean first of all the average conditions, computed by considering many different events.

We also mean the variability of weather phenomena around the average conditions: in other words, how intense and frequent are the weather variations around the average weather state. The climate statistics also include information on the frequency of some particular events, for example, the extremes that cause most of the weather-related damages.

When we talk about the events that affect our daily life—for example, whether it is sunny or rainy, or whether a storm is going to affect us in the next few days—we are talking about weather. We often hear people mentioning "the changing weather," referring to the fact that in a short time, even every hour, one can experience rather different conditions, one after the other, sunny periods followed by cloudy and windy ones, then rainy again, and finally back to sunny.

By contrast, when we talk about "the changing climate," we refer to the fact that the weather statistics—for example, the average weather conditions over a season, or a few years, and the variability around the average conditions—are changing.

Consider a town on the Tuscan hills and suppose that we want to describe its weather and climate, in terms of one typical weather variable, the surface temperature measured 2 meters above the ground. Figure 1.1 shows the minimum and maximum temperature during a summer and a winter month in the 1950s.

The daily cycle of temperature is evident, with a succession of warmer and colder values that characterize the day and night-time temperature (night temperatures are normally lower than day temperatures, due to the missing warming effect of the incoming solar radiation). If we consider a period of a few days, Figure 1.1 shows that weather changes, with relatively warmer days followed by relatively colder ones. Also note that during both months we have a few sequences of days with similar temperatures: if we are in a warm season, and these days are characterized by very high temperature, we talk about heat waves. Similarly, in winter, we talk about cold

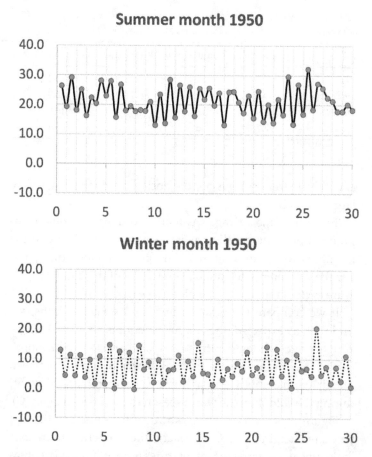

Figure 1.1. Minimum and maximum temperature (°C) in a summer and a winter month of 1950 at a town in Tuscany: each symbol represents a temperature value observed at a 12-hour interval.

spells when we detect a sequence of days with substantially lower-than-average values.

For each month, we can calculate the monthly-average temperature conditions and describe the daily changes in terms of the difference between the daily temperature and the monthly-average value. This difference (between the daily temperature and its monthly mean), is called the daily temperature

Table 1.1 Monthly-mean daily average, maximum and minimum temperature (°C), at a town in Tuscany, during a winter month and a summer month in 1950 and in 2020

	Winter 1950	Summer 1950	Winter 2020	Summer 2020
$<T>$	6.9	21.1	7.6	23.4
$<max\ T>$	10.1	24.2	11.2	25.3
$<min\ T>$	2.9	18.1	5.1	19.9
$<day\ cycle>$	7.2	6.1	6.1	5.4

anomaly. Extreme events are the ones characterized by a very large anomaly.

To describe the average weather conditions, we could compute not only the monthly average temperature, but also the monthly average maximum and minimum temperatures, and use their difference to gauge the intensity of the daily temperature cycle. Table 1.1 reports all these values.

If we compare the 1950 summer and winter months, we find that, as expected, the summer month is characterized by a higher average temperature (21.1°C) than the winter month (6.9°C). Note also that the daily cycle, the difference between the daily maximum and minimum temperature, is wider during the cold than the warm season (7.2°C instead of 6.1°C).

Figure 1.2 shows the temperature for the same months, but 70 years later, in 2020. At first sight, the temperatures in 2020 look rather similar to the temperatures in 1950. The daily values vary as before, and the summer month is warmer than the winter month. But if we compare the monthly average temperature and the monthly average daily variation (Table 1.1), we detect some differences. For both months, the average temperature is higher in 2020 than in 1950. Table 1.1 further indicates that the average of the minimum temperatures increased more than the average of the maximum temperatures. If we consider the summer months, the average minimum temperature increased by 1.8°C while the average maximum temperature increased by 1.2°C: the warming has been stronger during the

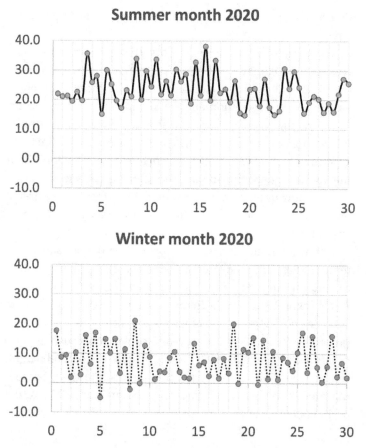

Figure 1.2. Minimum and maximum temperature (°C) in a summer and a winter month in 2020: each symbol represents a temperature value observed at a 12-hour interval.

night than during the day. We can also see that in 2020 the daily cycle is narrower, for both months: this is a consequence of the fact that the minimum temperatures have increased more than the maximum ones.

This analysis and the statistics reported in Table 1.1 tell us something about the climate, represented in this case by monthly-average weather conditions, and not about single weather events. From the values reported in Table 1.1, we can detect that the climate has changed, and that temperatures have become warmer.

This single example highlights the difference between weather and climate, looking in particular at surface air temperature, and with the climate defined by monthly average values. In the discussion about the climate, we limited ourselves to the average weather, and we have not looked at other statistics. A more complete comparison of the climate of 1950 and 2020 also would have included information about the variability around the average climate. This information would allow us to answer questions such as these: Are extreme weather events more or less frequent today? Has their intensity changed?

In this analysis we have also focused on one variable only, temperature. More complete and general climate reports consider a range of variables. For example, if we look at the Assessment Reports of the Intergovernmental Panel on Climate Change (IPCC), annual average values and decadal averages (averages computed over 10 years) of variables such as temperature at different levels in the atmosphere, sea-level height, sea-ice extension, and precipitation are used to describe how the climate has been changing since the preindustrial era (the period between 1850 and 1900).

1.2 Do weather and climate vary spatially and temporally?

Yes, both weather phenomena and climate statistics vary spatially and temporally.

While discussing the difference between weather and climate, we said that the climate describes the statistics of weather phenomena computed over a long time, with statistics represented, for example, by average temporal values. We have also discussed the fact that weather phenomena are characterized by frequent changes, by a frequent temporal variability. In other words, we have discussed the temporal scale of the weather phenomena and of climate.

Let us now introduce a second key aspect of weather phenomena that we need to consider when discussing weather and climate: the spatial scale of the weather and climate phenomena.

Weather is very sensitive to the spatial location. Locations that are spatially very close could experience very different weather. Consider, for example, two nearby towns, the first one on the coast and the second one a few kilometers inland. The effect of the ocean, the fact that it changes its temperature very slowly, over a long time (weeks and months rather than hours and days), has an impact on the temperature of the air masses that flow above it: as the ocean has a long-term thermal inertia, so do the low levels of the air flowing above it. As a result, the temperature variability of locations on coastal areas is rather small, and regions close to the ocean are less frequently affected by extreme conditions. By contrast, an inland town, especially if shaded by a mountain chain, would show a wider temperature variability. This can have a strong impact on the local weather that can be very different in the two towns. In general, this example indicates that the daily weather depends on the spatial location and can be very different if one moves to a nearby location.

Let us now consider both the spatial and temporal dimensions. Phenomena linked to moist processes are characterized by a very fine spatial and fast temporal scale. Think, for example, of a thunderstorm and how its position is influenced by the local terrain characteristics: as the wind blows over a complex terrain, the air can rise following the orography, and this can trigger, in certain conditions, the development of thunderstorms. As a result, the uphill side of a mountain could experience strong rainfall, while the downhill side dry and warmer conditions. Thunderstorms affect the weather not only on the spatial dimension, geographically, but also on the temporal dimension: a cumulus cloud can form over a few minutes and cause rain to start suddenly. Associated wind gusts can develop and affect the lower levels of the atmosphere and cause huge damages close to the ground. The whole phenomenon can last a few minutes, and then the system could dissipate and calmer weather conditions can return.

Thus, some weather phenomena can be very localized spatially and can change over a fast temporal scale (from a few minutes to a few hours): in physics we say that they are characterized by small spatial and fast temporal scales.

By contrast, the climate is characterized by longer temporal scales, since the climate is defined as the average weather, with averages computed over a long temporal window.

The climate is also characterized by large spatial scales. In fact, unless there is a substantial and profound difference between the geographical characteristics of two nearby locations, for example, if they are close but one is at the top and the other at the bottom of a steep mountain, their climate is usually very similar. This is why we usually talk about the climate of a region, rather than of a single location, where the region is chosen to include locations with similar characteristics, for example, in terms of altitude, terrain characteristics (forest, grassland, or city), or position with respect to the ocean or a large water basin.

1.3 Is there a clear separation between weather and climate?

No, there is not a clear demarcation line that separates weather and climate.

If we characterize phenomena by their spatial and temporal characteristics, weather events span the whole spectrum. We associate fast and small-scale phenomena with very localized weather events (for example, moist processes) and slow and large-scale phenomena to weather events that affect a large region, a continent, or a large part of an ocean, and that evolve slowly, over weeks or months. Some large-scale weather conditions, such as a high-pressure system that affects the

North-Eastern Atlantic region or an El Niño event that affects the tropical Pacific, evolve slowly. Thus, weather includes both small spatial and fast temporal scales, and large spatial and slow temporal scales.

When we study the climate, we are concerned about the statistics of weather events, computed considering a long period of time, at least a month but ideally a few years. These statistics include the statistics of the small spatial and fast temporal scales (e.g., the average intensity and scale of winter storms) and of large spatial and slow temporal scales.

Thus, both weather and climate include small spatial and fast temporal phenomena, and large spatial and slow temporal phenomena. We cannot say that climate is characterized only by spatial scales larger than a certain spatial threshold or by temporal scales slower than a certain temporal threshold, since the climate is also defined in terms of the statistics of the small spatial and fast temporal scales.

1.4 Is weather affected by all (fast and small-scale and slow and large-scale) phenomena?

Yes, all scales contribute.

Any phenomenon that affects a physical system can be characterized by its spatial and temporal characteristic scales. Complex motions can be broken down and described in terms of interacting waves, and this applies also to the weather. Thus, the simplest weather phenomenon can be described as a wave that evolves and moves (in physics we say "propagates"): a wave with a specific spatial characteristic, its wavelength, and time characteristic, the period that it takes for the wave to cross a fixed reference point. Waves similar to the ones that propagate on the surface of the sea propagate in three-dimensional fluids, such as the atmosphere or the ocean. Waves develop at a certain "initial" time, evolve for a certain amount of time while interacting with other waves with both a faster and smaller scale and a slower and larger scale, and then decay.

An example of a complex system determined by the interaction of many single waves is the surface of the ocean close to a shore. We can describe it as a combination of many waves with different characteristics. Some have more energy than others and propagate further ashore, while others are slower and gentler, and remain offshore. The sea state includes many of these waves, all interacting, and the sea surface and its roughness (the degree of small-scale details that it includes) are determined by their interaction.

Another example is the waves that you can create on a rope, after you have fixed one end on a hook and you hold and move the other end in your hand. Depending on how fast you move your hand, and how wide your movement is, you can generate waves with different characteristics. If you start moving your hand randomly, without following any specific pattern, you could generate waves with a wide range of spatial and temporal characteristics. As your hand moves, it transfers energy to the rope, which oscillates, generating a "spectrum" of waves.

Similarly, the atmosphere and the ocean are three-dimensional fluids that include a wide range of interacting waves. At any time, waves with different spatial and temporal scales coexist. Depending on their characteristics, some waves extract energy from the others and thus grow in scale and amplitude up to when they start giving back part of their energy to the other waves, and start diminishing their strength and eventually disappear. A hurricane is an example of such a wave.

Have you ever looked at an animation of a satellite image? Or the animation of a weather forecast of cloudiness? If you do, you will see a continuous range (in physics, we talk about a wave spectrum) of waves interacting, propagating, growing, and decaying, as they exchange energy and other properties.

Does the fact that the atmosphere wave spectrum is continuous have any implications for numerical weather prediction and climate projection? Yes, it does: it implies that if we want to build a realistic model of the atmosphere, or of the ocean, we need to represent all of them and simulate how they interact. If, instead, the model that we are using to describe the atmosphere has some approximations (e.g., a finite spatial resolution) and cannot simulate waves with a spatial scale finer than a certain resolution, we will not be able to simulate the effect that the unrepresented waves have on the represented waves. A model defined on a finite mesh cannot simulate waves finer than its resolution, and it has difficulties in simulating in a realistic way their impact on the simulated large-scale waves. Missing or misrepresenting these scales can have a substantial impact on the model capability to realistically simulate the atmosphere. The ocean is similar to the atmosphere: its wave spectrum is continuous.

1.5 Which coordinate system is used to study atmospheric and oceanic motions?

Global weather and climate phenomena are studied on a rotating, spherical coordinate system.

Let me remind you that to characterize the position of an object in space, we need three coordinates: most typically, for systems where gravity plays a role, two coordinates describe the position of the object on a horizontal plane perpendicular to the gravitational force, and one describes its distance in a direction perpendicular to this plane. A typical example of such a coordinate system is the Cartesian one, with the three directions orthogonal to each other. Once the coordinate system has been chosen, velocities are defined as the rate of change of the position along these three directions.

It is also worth recalling three important characteristics of the global motions in the atmosphere or in the ocean: the Earth is a sphere, the sphere rotates, and the depth of the two fluids (the atmosphere and the ocean) is small compared to the radius of the sphere. These three facts are important and determine the coordinate system used to study atmospheric and oceanic motions:

- The fact that the Earth is a sphere imposes that we use a spherical coordinate system, whereby distances are measured in terms of the distance from the center of the sphere, and two angles, the latitude and the longitude. These two angles are often replaced by the distance from the Greenwich meridian along a latitudinal circle, and by the distance north or south of the equator.
- The fact that the Earth is a rotating sphere means that the coordinate system that we are using continuously accelerates, and this has to be considered when deriving the equation of motions. The main effect of this fact is that the equation of motion written with respect to a rotating system of reference includes some extra terms (extra in the sense that they appear in the equation only because the system of reference accelerates). These extra terms needs to be considered.
- The fact that the two fluids are thin compared to the dimension of the Earth means that some terms of the equations of motions can be simplified.

Because of these three facts, the coordinate system that is used to study global atmospheric and oceanic motions is a rotating spherical coordinate system, with the three coordinates being the latitude (i.e., the angle measured with respect to the equator), the longitude (i.e., the angle measured with respect to the Greenwich meridian), and the height (measured from sea level).

This coordinate system can be simplified back to a Cartesian coordinate system when one is interested in local motions, spanning only a limited geographical region and occurring over on a short time.

1.6 What are the key weather variables?

The key weather variables that are mostly used to describe the weather are temperature, wind, pressure, and humidity.

A few key variables are normally used to describe a weather phenomenon and to build realistic models of the atmosphere. Since the climate is defined as the average weather, the same variables can be used to describe the climate. These variables are as follows:

- *Temperature*: expressed in degrees Kelvin or Celsius, it is a measure of the atmosphere or the ocean internal energy, of how fast its molecules vibrate and oscillate; in general, for any gas temperature is a measure of its internal energy.
- *Velocity*: expressed in meters per second, it measures how fast masses of air move around. Velocity is a three-dimensional vector, with a component along each of the three axes of motion (the two horizontal coordinates, normally chosen along the west-to-east and the south-to-north directions; and the vertical coordinate, normally chosen vertical to the Earth's surface).
- *Specific humidity*: expressed in terms of grams of water per kilogram of air, it is a measure of the mass of water vapor per unit of air.
- *Pressure*: expressed in pascals (i.e., in newtons per square meter), it is a measure of the force per unit area.

At any location and specific time, the state of the atmosphere (the weather) can be described in terms of these four variables. By applying the laws of physics to the atmosphere, written in terms of these variables, we can deduce a set of equations that describe how the atmosphere evolves in a realistic way. In other words, these variables can be sufficient to capture the key features of the actual weather, and can thus also be used to characterize its climate.

Other secondary variables could be computed from these four, or new variables could be introduced to provide more details of the state of the atmosphere. For example, one might want to define a variable that describes the sky cloudiness: this variable can be expressed in terms of the percentage of the sky covered by clouds. To describe more accurately the moist processes that occur within a cloud, one might want to introduce variables that describe the concentration of liquid water, snow, and ice crystals within a cloud. A person who lives close to the sea might want to describe the weather also in terms of the sea state: this could be described by the ocean surface wave heights, direction, and frequency. As models become more accurate and complex, and aim to describe also how the chemical components of the atmosphere evolve (think, e.g., of the carbon cycle and how CO_2 is transported and absorbed by vegetation or the oceans), more variables have been, are, and will be introduced.

1.7 Why does weather change?

Weather changes due to the continuous exchange of energy and momentum between air masses and between the air and the land and ocean surfaces.

The local weather, the weather that we experience every day, is characterized by a succession of changes, some of them happening very rapidly and others happening at a slower pace. The atmosphere is a three-dimensional fluid: thus, the local weather is determined by the three-dimensional interaction between the different waves that propagate in the atmosphere and by the interaction of the air masses with the Earth's surface. The local, surface (i.e., close to the ground) weather is determined by the interaction between the upper-level atmospheric flow, which evolves a few hundred-to-thousand meters above the Earth's surface, and the local features of the Earth's surface.

The local weather is also determined by the complex interactions between the upper-level, larger-scale flow, which usually varies on a rather slow timescale (say hours and days) and the lower-level flow, which varies on a rather faster time-scale (say minutes and hours); it includes the small-scale waves that are generated when the lower-level flow interacts with the Earth's surface. These interactions include, for example, air flowing on top of ocean waves or among the buildings of a city, the trees of a forest, the hills, and the mountains.

1.8 How are motions in the atmosphere generated?

The sun is the source of the energy that generates the waves that propagate in the atmosphere, and thus the weather phenomena.

Geographical differences between the amount of solar energy hitting the atmosphere and the Earth's surface determine differences in the air temperature: for example, one area could be covered by more clouds than another one, and thus could reflect a larger amount of incoming radiation. These differences cause variations in the temperature of the air masses. Differences in the temperature of air masses lead to differences in pressure, and differences in pressure (let us remember that pressure is defined as the force that acts on a surface with a unit area) cause air masses to move.

A further complication arises from the fact that the Earth rotates, and we live on this rotating frame of reference: this fact needs to be considered when we study atmospheric motions, and when we deduce the equations to understand and predict them. Also, let us not forget gravity, which acts to attract any mass of air toward the center of the Earth. And finally let us not forget friction, which causes the lowest levels of the atmosphere that flow close to the surface to decelerate and lose energy. The sum—the resultant of the pressure force, the force of gravity, the Earth's rotation and friction—determines the motions.

As the air masses start to circulate, the interaction of the flow with the land surface generates disturbances, which can be represented as a combination of waves. Think, for example, of the westerly flow, which goes from the west toward the east, crossing the Pacific, and then impinging onto the Rocky Mountains. The low-level flow, the flow close to the surface, is forced to go either around or above the mountains. Even where the terrain is rather flat, when the flow crosses the land-sea boundary, it is disturbed by the fact that the two surfaces, land and sea, have different temperature and roughness, and thus needs to adjust. Over the sea, the low-level atmospheric flow exchanges heat and momentum with the sea: the wind causes the ocean waves to grow, and the ocean waves generate a frictional force that slows the wind down. More generally, as air flows above any surface, frictional forces slow it down and induce small-scale waves in the air, which in turn interact with the larger-scale waves and affect them.

In summer, when moist air masses flowing over a warm sea hit a land mass with a complex terrain, they are forced to di-verge and move around the obstacle, and/or to rise. As they rise and encounter regions of colder air, they start to cool. As they cool, moisture starts condensing, generating clouds and pos-sibly precipitation. This phenomenon can be very localized, or it could lead to more widespread rain if the upper-level atmos-pheric flow favors the formation of organized cloud systems that can eventually grow in scale into clusters of convection clouds affecting large regions. If this is the case, we could have a large-scale area characterized by similar, convectively in-duced weather events. Over some ocean regions, for example, convection can self-organize over large pools of warm sea water, develop, intensify, and then propagate.

A single convective cell can be considered as a small-scale wave, with a characteristic length scale of few hundred meters and a characteristics temporal scale of between a few minutes and a few hours. This means that the cell would affect an area of a few hundred meters, and that in a short time period moisture would have condensed and precipitated, and then the cell would have dissolved.

Organized convection (compounded by close-by, interactive convective cells) can induce changes in the atmospheric flow and trigger the development and evolution of large-scale upper-level waves, with a characteristic length scale of few thousand kilometers and a temporal scale of days. Organized convection could propagate and affect other areas, and in its path continue to develop and send large-scale waves toward the upper levels, both toward the northern and the southern latitudes.

A small-scale convective system can develop in the tropical region, grow and transform itself into a larger-scale, organized convection system, and affect the atmospheric circulation far away, into the northern and southern latitudes. Seen from a satellite, this evolution would appear as a cluster of interactive small-scale waves that eventually merge and evolve into a large-scale phenomenon.

In winter, over the extra tropics (the Earth's surface outside the tropical band between the latitudes of 30°S and 30°N), we tend to see more organized and larger-scale systems than during the summer. These large-scale systems have a characteristic length scale of a few hundred kilometers: they are called "synoptic-scale" systems, and include phenomena such as a deep cyclonic circulation or an area of high pressure. They are more frequent in winter than in summer because of the larger difference in heat between the lower and higher latitudes, which induces stronger winds and more organized and energetic systems, which develop, evolve, and decay over a period of few days. These systems are defined by interacting large-scale waves, with a characteristic length scale of kilometers and a characteristic temporal scale of days.

The most typical synoptic features that are usually highlighted on weather maps and are used to describe weather forecasts are high- and low-pressure systems. In the Northern Hemisphere, high-pressure systems, also called anticyclones, are characterized by a clockwise circulation, while low-pressure, cyclonic systems are characterized by an anticlockwise circulation.

The most energetic waves that develop and propagate in the atmosphere are tropical storms, which can intensify to become hurricanes and typhoons. Let us consider, for example, a tropical storm crossing the Atlantic Ocean: it is usually born as small-scale waves off the western coast of Africa. It then travels embedded in the large-scale flow, which in that region goes from the east to the west. As it travels over the warm, tropical Atlantic Ocean, it extracts energy from the ocean and the warm and moist low-level air, and grows in size and strength. It keeps propagating and strengthening until it cannot extract any more energy from the surroundings, for example, because it has made landfall and cannot extract heat and moisture from the underlying warm ocean, or because it has moved over ocean areas with colder temperature.

Large-scale geographical features such as land-sea contrasts and mountain chains (e.g., the Rockies, the Andes, the Alps, and the Himalayas) can force the atmospheric flow either to diverge or to rise and can induce large-scale, slowly evolving perturbations to the flow. Differences in soil characteristics, for example, linked to the presence of deserts, large forests, or water basins, can induce large-scale, slowly evolving perturbations to the flow that can grow into features that can last for days, weeks, and possibly months. These features are usually referred to as low-frequency phenomena. Examples are large regions of high pressure, with a scale of a few thousand kilometers, surrounded to the west and the east by areas of low pressure. When the low- and high-pressure regions are well established, the atmospheric flow sometimes follows a well-defined omega-like circulation that can affect a large part of a hemisphere. These features are called "blocks," or "omega-blocks" when an omega feature is evident, to highlight the fact that the atmospheric flow has to move around the high-pressure area, which blocks it from flowing in a straight line.

Europe and the Euro-Atlantic sectors are regions that are often affected by these blocked circulations and other low-frequency phenomena. One of these low-frequency phenomena is the North Atlantic Oscillation (NAO), a pattern that determines the flow configuration over the northeastern Atlantic Ocean and Northern and Central Europe. Similar large-scale, low-frequency phenomena can be identified in other regions of the world: for example, the Pacific North America (PNA) pattern over the northeastern Pacific and Northern America, or the Arctic Oscillation (AO).

These large-scale patterns that characterize the flow over almost a quarter of a hemisphere last longer than smaller-scale features such as high- and low-pressure synoptic systems. They can last from several days to even a few weeks, and during this period can be quasi stationary (i.e., they hardly propagate), and as a consequence the weather at the surface does not change.

1.9 Key points discussed in Chapter 1 "Weather and climate"

These are the key points discussed in this chapter:

- With the term *weather* we mean the phenomena that we experience every day, phenomena due to changes in the state of the atmosphere, and/or of the Earth system components (atmosphere, oceans, sea ice, land) that affect the weather.
- Weather phenomena are characterized by their spatial and temporal characteristics; they can be described by waves that develop, grow, interact with the other waves, and eventually decay.
- The atmosphere includes waves of all scales that interact; the range (i.e., the spectrum) of these waves is a continuum.
- With the term *climate* we mean the statistics of the weather, where the statistics are computed over a long

time, say a few months or years; the simplest statistical property that is usually computed to define the climate is the average weather.

- There is not a clear separation between weather and climate: annual-mean weather forecasts blend seamlessly with decadal and multidecadal (climate) forecasts.
- Weather and climate are affected by all scales.
- Earth system motions are studied in a rotating, spherical coordinate system, with positions defined in terms of the latitude, longitude, and above-sea-level altitude.
- The few, key weather variables that are normally used to describe weather phenomena and the climate are temperature, wind, pressure, and humidity.
- Atmospheric motions are caused by the interaction of air masses with different characteristics (temperature, density, humidity, and pressure), subjected to the incoming solar radiation and the outgoing long-wave radiation, the action of gravity and friction, and of the fictitious force of Coriolis, due to the fact that the coordinate system rotates.

2

THE EARTH SYSTEM

In this chapter we describe the components of the Earth system that determine the weather phenomena, introduce the notion of an accurate and skillful forecast, and discuss how heat is transported and exchanged and where the energy that drives weather phenomena comes from. More specifically, we will be addressing the following questions:

1. What does "Earth system" mean in numerical weather prediction?
2. What is an accurate and skillful forecast?
3. What are the key building blocks of an Earth system model?
4. What are the key processes simulated by an Earth system model?
5. How is heat transported and exchanged in the Earth system?
6. Where does the energy that drives the Earth's climate come from?
7. What are the key similarities and differences of the atmosphere and the ocean?

2.1 What does "Earth system" mean in numerical weather prediction?

In numerical weather prediction, with the term "Earth system" we mean the essential units and processes of the planet's physical system that determine and influence weather phenomena.

In the early days of numerical weather prediction, the models that were used operationally to generate the forecasts included the atmosphere and a very simple model of the land-surface processes. They did not include a dynamical ocean, did not simulate the interaction of the ocean waves with the lowest layer of the atmosphere, and considered sea ice as static. This limitation was, in some way, a necessary choice: the models had to be kept simple, since the available computer power was very limited compared to today, and it was impossible to integrate in time a complex model that included more physical systems to generate a weather forecast. Furthermore, the role of some slowly evolving components on the forecasts that were issued at that time, which extended only up to 7 to 10 days, was not evident and/or was underestimated. The approach was to include the processes that were dominant: the processes that, if neglected, would have caused forecasts to behave very differently from reality during the forecast range.

Fortunately, since the early 1980s, computer power availability has been increasing by about a factor of 2 every 18 to 24 months, in the sense that for the same amount of money, every 18 to 24 months we could have access to double the computer power that we had. During this period, national meteorological centers changed their supercomputers about every 4 years, with an upgrade at the midterm of their contract. In this way, every 4 years they had access to ~5 times more computer power. Over 40 years, this implies an increase in computer power of about a factor of 10^6–10^7.

As more computer power became available, models could be made more complex: the simulation of the processes that were already inserted in the models could be made more realistic, and missing processes could be included. During the same period, more and better observations became available, allowing model developers to diagnose more thoroughly model performance and improve them. As models progressed, more observations could be used to initialize the forecasts, and as this made forecasts more accurate, scientists decided to extend the forecast length beyond the 7-to-10-day forecast (used until the beginning of the 1990s). As the forecast length was extended, it became evident that more processes had to be included in the models to improve their realism beyond a few forecast days. In particular, we realized that we needed to include the slow processes that controlled the evolution of the large-scale waves. This is why dynamical ocean, dynamical sea-ice, and improved land-surface models were gradually included in the numerical weather prediction models.

Today, an Earth system model includes the simulation of processes in the atmosphere and the ocean, the sea ice and vegetation, and deep-soil processes. Thanks to model advances and the shift from atmosphere-land-only to Earth system models, the availability of more and better observations, and the adoption of new forecasting approaches, the more advanced weather prediction centers can issue accurate and skillful forecasts valid for 1 day to a few months ahead.

2.2 What is an accurate and skillful forecast?

A forecast is labeled accurate if it is close to reality, and it is labeled skillful if it is closer to reality than a reference forecast, for example, defined by the past observed statistics.

To answer this question, let us first illustrate how we assess whether our model is accurate and represents in a realistic way all the key processes that determine the weather, and then let us define the concept of a "skillful forecast."

Forecast accuracy is usually assessed by both analysing the quality of individual model components in simulating "single processes," and by looking at the overall performance of the "entire model."

Let us start by focusing on some specific phenomena, for example, moist processes. To assess whether a model is accurate, we usually select a few key locations, where moist processes play a key role in determining the weather (e.g., in the tropics and in the extra-tropics, close to the ocean, at the top of a mountain and at the bottom of a hill, in a forest), and install in each of them a set of instruments that measure the key variables that characterize moist processes (e.g., temperature, humidity, rain). We then generate our model simulations and check whether when rain is observed, rain is also simulated in the model. We use the detailed observations given by the installed instruments to check whether the model is capable of simulating how an air mass becomes saturated with moisture, condensation starts, and then precipitation is initiated. We compare how temperature and humidity evolve in reality and in the model. At the end of this comparison, we can draw robust conclusions on whether the model component designed to simulate moist processes is realistic and "good enough" for our needs. By focusing on specific processes and looking at a few specific locations, we can perform a thorough investigation of how the model simulates specific weather events.

Once this first set of "single process" testing has been completed, we need to assess how the whole model performs in a variety of situations, to avoid the risk of building a model that works perfectly well only in some specific weather conditions and/or locations. This second phase is necessary to assess whether the interaction between the processes in the model matches their interactions in reality, everywhere and in any condition. For example, once we have tested independently the module that simulates moist processes and the module that simulates the impact of mountains on the atmospheric flow, we need to assess whether, when the two schemes are used together, they still produce realistic simulations in all locations, and in any weather condition.

One way to assess the "entire model" accuracy is to use it to generate weather forecasts. Later in the book we will discuss in detail how one can use a model to generate a forecast. For now, let us just assume that we can generate a forecast valid for the next few days starting from the current state of the atmosphere. We can compare our forecasts with what happens in reality and measure how close or distant our forecast is from reality.

The distance between our forecasts and reality is a measure of the model error: the longer the distance, the further away we are from reality and the larger the error is. As a measure of forecast error over a region (e.g., the United States, or Europe, or a whole hemisphere), we could use the average distance between our forecast and reality, computed by averaging the distance between the forecast and reality within the verification region. Once we have computed the average error over a region for one forecast, we can repeat the process considering forecasts valid for different days, and thus have a more robust estimation of forecast accuracy based on many different weather forecasts.

Once we have completed this assessment for a whole season (say 90–100 days), we can draw conclusions about the accuracy of our model. For example, we can analyze how the forecast error evolves with time and determine how long it takes for the initial error to double, or to increase by a factor of 10. This type of assessment is performed routinely, to have an up-to-date forecast evaluation, and to monitor our long-term progress in numerical weather prediction.

The forecast error evolves as the forecast time progresses: it stays rather small for very short forecast times (say, for forecasts valid for a few days) and then grows until it saturates and reaches an asymptotic value. Depending on the variable and the area that we consider, the short-range forecast error can be very small, or already significant: this could happen, for example, if we consider variables that are very sensitive to local details, or locations where the model performs very poorly. As the forecast time progresses, also the speed at which the error reaches the asymptotic value depends on the variable and the region. For example, the error of the forecast of the average temperature over a large area such as Europe reaches the asymptotic level in about 2 weeks, while the error of the forecast of cloud cover reaches its asymptotic value already at about forecast day 3 to 4.

Now that we know how to assess the forecast accuracy, we can define the forecast skill as the accuracy of our forecast compared to a reference forecast. The reference could be, for example, persistence, whereby the weather of the forthcoming days is assumed to be equal to today's weather, or climatology, whereby the weather of the forthcoming days is set to be the average climate conditions observed in the past decades.

Determining whether a model is skillful with respect to reference benchmarks is very important, since it gives a measure of the gains that using a forecast can bring, and of the return of the investments put into building the model and generating the forecast.

2.3 What are the key building blocks of an Earth system model?

The key building blocks of an Earth system model are lines of software capable of simulating in a realistic way the most relevant processes of the Earth system.

Now that we have defined the concept of a skillful forecast, let us resume our discussion on Earth system modeling. We were saying that, in the early 1980s, forecast models were rather simple, and included only atmospheric processes and a very crude representation of land-surface processes. Then, year after year, as models improved, the skill was extended: for example, the error of a 3-day forecast in 1990 decreased to the level of the 2-day forecast in 1980, indicating a gain in forecast skill of 1 day in a decade. Forecasts of the upper-level, large-scale flow, i.e., of the waves that characterize the atmospheric flow at an altitude of about 5,000 meters, started being skillful up to about 1 week, and people began wondering whether and how the forecast skill could be extended even further.

Evidence started to emerge that slowly evolving components of the Earth system that were not included in the weather models had an influence on the local weather, and thus if they were included could help in improving the forecast skill. For example, patterns of sea surface temperature variations due to ocean currents were shown to have an impact on the organization of tropical convection, and the formation of tropical large-scale organized convection patterns were shown to influence the weather of the extra-tropics.

Numerical experiments also showed that large-scale changes in the land moisture content were having a substantial impact on the establishment and intensification of extra-tropical, large-spatial-scale, high-pressure systems, especially during summer. Other experiments suggested that the simple land-surface schemes developed in the 1980s needed to be expanded to include more soil layers, so that they could represent in a more accurate way the fluxes of moisture and heat between the surface of the Earth and the subsoil layers over longer time ranges (say weeks and months, instead of only hours and days), and how these fluxes depended on the soil characteristics (desert, grassland, forest, cities). Dynamical vegetation modules capable of simulating how plants evolve in time during a forecast started being tested and included. Other sets of numerical experiments indicated that we needed to include more sophisticated ocean models, capable of simulating realistically both the ocean waves and the three-dimensional currents, and that we needed to include dynamical sea-ice models able to react to changes both of the atmosphere (e.g., temperature) and the ocean (e.g., interaction with the waves) conditions.

These developments transformed the characteristics of numerical weather prediction models, from being essentially a model of the atmosphere with a simple description of the land-surface processes, to complex models that included dynamical simulations of atmospheric, land-surface, ocean, and sea-ice processes. These complex Earth system models were first used to study the climate, since we realized that coupling was essential if one wanted to generate long-term forecasts spanning years. But then were adopted also in weather prediction as we realized that forecasts covering the forecast range from 2 weeks to a few months would benefit from using Earth system models. More recently, we had evidence that medium-range forecasts of hurricanes generated by Earth system models that included a dynamical ocean and sea-waves were better. Indeed, today at the most advanced research and operational centers, Earth system models are used both for weather and climate studies.

2.4 What are the key processes simulated by an Earth system model?

There are many key processes that we need to simulate if we want to generate accurate and skillful forecasts: generally speaking, their number and complexity increase the finer the details we want to predict and/or the longer the time range we want to cover.

To answer this question, let us list the processes simulated at the time of writing (summer 2022) by the global model of the European Centre for Medium-Range Weather Forecasts (ECMWF), one of the most realistic global models that is used daily to generate forecasts valid from 1 day to 1 year, and is also used by the EC-Earth Consortium to generate climate simulations (see also a schematic in Figure 2.1):

Atmosphere:

- *Radiation*: to consider the impact on air masses of the incoming solar short-wave radiation and the outgoing long-wave radiation
- *Turbulence*: to simulate the impact of the interaction of waves with different scales from each other

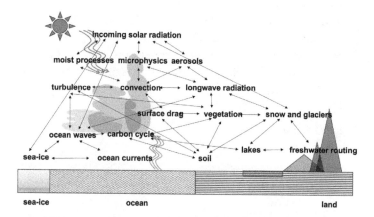

Figure 2.1. Schematic of the key processes simulated in an Earth system model and their dominant interactions.

- *Turbulence, energy, and momentum diffusion*: to stabilize the numerical integration procedures and simulate the impact of sub-grid-scale processes onto the scales that are resolved by the model
- *Gravity*: to consider its impact on air masses with different density
- *Coriolis force*: to consider the fact that the Earth rotates, and we predict the weather with respect to a rotating frame of reference
- *Moist processes and water cycle*: to simulate the water cycle, generate precipitation, and simulate the impact of moist processes on the fluxes of energy and momentum between different air masses
- *Microphysics*: to simulate how moisture condenses and is transformed from water to snow and ice droplets, and vice versa, and to transport these different meteors and make them interact with radiation
- *Sub-grid-scale processes*: to account for the processes that a model cannot represent explicitly because of the finite resolution of its grid

- *Carbon cycle*: to simulate its impact on vegetation and on energy transfer in the atmosphere
- *Aerosols*: to simulate their impact on radiation and to be able to predict how they propagate in the atmospheric flow

Land surface:

- *Surface drag and turbulence*: to consider how different surfaces impact and slow down the low-level atmospheric flow
- *Vegetation*: to consider the impact of plants on the radiation fluxes and on the transport of moisture from the soil to the atmosphere
- *Orographic forcing*: to simulate the impact of complex terrains on the low-level atmospheric flow
- *Soil heat and moisture fluxes*: to consider how moisture and heat are transferred throughout different soil types
- *Freshwater routing*: to take into account how precipitated water is channeled and transported over land
- *Snow and glaciers*: to simulate their impact on radiation and the low-level atmospheric flow
- *Lakes*: to consider their effect on the (mainly low-level) air masses
- *Cities*: to be able to simulate their effect on radiation and the low-level air characteristics (temperature and velocity)

Ocean:

- *Waves*: to consider their effect on the lowest layer of the atmosphere, and on sea ice and the coastal areas
- *Three-dimensional currents*: to be able to distribute correctly heat in the ocean, by moving water masses with different characteristics, and to simulate their impact on the ocean waves and indirectly on the atmosphere

- *Ocean convection*: to simulate vertical motions in the ocean, due to differences in density and temperature of water masses
- *Ocean turbulence*: to simulate the impact of the interaction of waves with different scales onto each other
- *Diffusion*: to stabilize the numerical integration procedures

Sea ice:

- *Interaction of sea ice with the atmospheric low-level flow*: to consider the impact of sea ice on radiation, and on the temperature and wind characteristics of the lowest layer of atmosphere
- *Interaction of sea ice with ocean waves and currents*: to simulate the two-way interaction between the ocean (waves and currents) and sea ice, and to simulate the ocean wave-breaking effect on sea ice

By including and making as realistic as possible the simulation of these processes, weather forecasts have improved by about 1 to 2 days every 10 years. Thanks to these advances, a 7-to-10-day forecast issued in 2020 has the same accuracy as a 2-to-3-day forecast that was issued in the 1980s. Because of this remarkable increase in skill, today weather-risk management decisions can be taken days before events happen, thus leading to better preparations to face extreme events and reducing their potential harm.

2.5 How is heat transported and exchanged in the Earth system?

Heat is transported via radiation, conduction, and convection.

The Earth is a physical system, and thus it obeys the laws of physics. In particular, it obeys the first law of thermodynamics that states that energy is conserved. When energy passes, as work or as heat, through matter (e.g., the atmosphere, or the ocean), the system's internal energy changes so that the total energy is conserved.

If we consider the atmosphere, we can measure its internal energy by its temperature (as is the case for any gas). Thus, the first law of thermodynamics says that if energy passes through a volume of the atmosphere, it will either change its temperature, or it will trigger a volume expansion or contraction. These two latter phenomena cause the volume of air to do work against the nearby volumes of air.

Heat can be transported to and from a system in three ways:

- *Radiation*: when no mass is exchanged, and no medium is required.
- *Conduction*: in this case, a medium (air, water) transfers heat by collisions between molecules, without any mass exchange.
- *Convection*: in this case, mass is exchanged, although no net real movement of mass may occur if masses of air with different energy simply exchange places.

Any volume of the atmosphere behaves as a physical body and reaches an equilibrium temperature that is determined by the application of the first law of thermodynamics. This temperature makes the sum of the internal energy and of the work that the volume is doing against any nearby volumes of the atmosphere to balance the incoming energy.

2.6 Where does the energy that drives the Earth's climate come from?

The primary source of energy for the Earth system is the sun, which outputs a "quasi" constant flux of energy toward the Earth.

Knowing where the energy that drives the Earth system comes from and how it is distributed among the Earth system components is essential to understanding weather events and how the climate evolves.

The sun behaves as any physical body with a temperature above zero degree Kelvin, and it emits a certain amount of radiation with a characteristic that depends on its temperature (more precisely, the relationship between the temperature of a "black" body, an idealized body that absorbs all the incoming radiation, and the radiation it emits follows the Stefan-Boltzmann law of physics). Because it is very hot, with a temperature of about 5,796 K (degree Kelvin), the sun emits a large amount of radiation with a very short wavelength. Note that the Earth itself is a physical body that emits radiation, but being colder than the sun, it emits a smaller amount of radiation than the sun, with a longer wavelength.

The radiation emitted by the sun hits the side of the Earth that faces it, passes through the atmosphere, is partly absorbed and partly reflected, and eventually reaches the surface of the Earth. To take into account the fact that part of the incoming solar radiation is reflected back into space, we introduce the concept of the albedo of the Earth surface, which represents the ratio between the reflected and the incoming radiation. The albedo depends on the surface characteristics: clean "white" sea ice, for example, has an average albedo of 0.84 (i.e., 84% of the incoming solar radiation is reflected back to space, and only 16% is absorbed by sea ice), while a green forest has an average albedo of about 0.14. Using observations from satellites, it has been estimated that the annual global average albedo of the Earth is 0.30, which means that globally and on average over a year, about 30% of the incoming solar radiation is reflected back into space.

The incoming solar radiation is the source of energy for the Earth system: part of it is reflected back into space (by clouds, aerosols, the surface), and part of it is absorbed by the different Earth system components, the atmosphere, and the land surface (the ocean, the cryosphere, and the land). As the Earth's components absorb it, they increase their temperature, and they themselves act as any physical body and emit radiation. Since they have a much lower temperature than the sun, their emitted radiation has a lower intensity and a longer wavelength: because of this characteristic, this long-wave radiation is also named infrared radiation. The infrared radiation emitted by the land surface is partly absorbed by the atmosphere and partly flows to the outer space.

The atmosphere itself acts as a black body: it absorbs the incoming solar radiation and the infrared radiation emitted by the Earth surface, warms until it reaches its equilibrium temperature, and then emits as a physical body at that temperature. Given that its temperature is closer to the temperature of the Earth surface than the one of the sun, its emitted radiation characteristics are similar to the infrared radiation emitted by the land surface.

Figures 2.2–2.5 are schematics of the energy fluxes between the sun, the atmosphere, and the land surface (data and schematics from Hartmann 2006). Values are expressed in watt/m^2, that is, in terms of the amount of energy per second expressed in watts that hits 1 squared meter, with 1 watt being equal to 1 joule per second (J/s), where the joule is a measure of energy. As a reference value for the energy, a light bulb of 100 watts, that is, of 100 J/s, consumes 100 joules of energy every second it is switched on.

Let us start by looking the incoming solar radiation (Figure 2.2): 340 W/m^2 arrive at the top of the atmosphere:

- 80 W/m^2 are absorbed by the atmosphere.
- 160 W/m^2 are absorbed by the surface.
- 100 W/m^2 are reflected back into space.

Now let us consider the radiation emitted by the Earth surface (Figure 2.3): 396 W/m^2 are emitted from the surface toward the atmosphere and the free space:

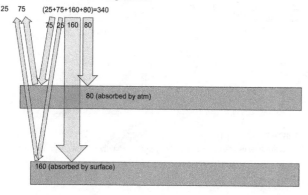

Figure 2.2. Energy fluxes of the incoming solar radiation: 340 W/m² arrive at the top of the atmosphere: of them, 80 W/m² are absorbed by the atmosphere, 160 W/m² are absorbed by the Earth's surface, and 100 W/m² are reflected back into space. (Source: Hartmann 2006)

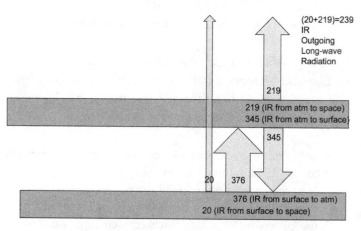

Figure 2.3. Energy fluxes of the infrared long-wave radiation emitted by the Earth's surface: 396 W/m² are emitted, and of them 376 W/m² are absorbed by the atmosphere and 20 W/m² go out into space. Of the 376 W/m² that the atmosphere absorbs, 345 W/m² are re-emitted toward the surface and 219 W/m² are emitted into space. Thus, in total, 239 W/m² are emitted back into space. (Source: Hartmann 2006)

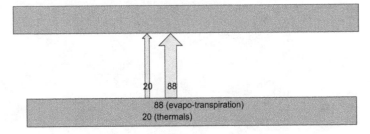

Figure 2.4. Energy fluxes from the Earth's surface: 108 W/m² are emitted, and of them 20 W/m² flow as thermals (masses of warm air that move from close to the surface to the upper layers of the atmosphere), and 88 W/m² are used in evapo-transpiration processes. (Source: Hartmann 2006)

- 20 W/m² go out into space.
- 376 W/m² are absorbed by the atmosphere; of them:
  - 345 W/m² are re-emitted toward the surface; and
  - 219 W/m² are emitted into space.
- Thus, in total, 239 W/m² are emitted back into space.

Finally, let us consider the radiation emitted by the land surface and transported to the atmosphere by convection and evapo-transpiration processes (Figure 2.4): 100 W/m² are emitted from the surface toward the atmosphere:

- 20 W/m² are associated with mass movements (thermals).
- 88 W/m² are linked to evapo-transpiration processes (e.g., an air mass close to the surface and rich with water vapor starts rising; as it reaches cooler layers of the atmosphere, water vapor starts condensing, and thus releasing energy).

Figure 2.5 brings all the energy fluxes shown in Figures 2.2–2.4 together. Note that there is a +0.6 W/m² net difference between the incoming solar radiation (340 W/m²), the reflected solar radiation (–100 W/m²), and the outgoing infrared (–239 W/m²): this +0.6 W/m² is an estimate of the

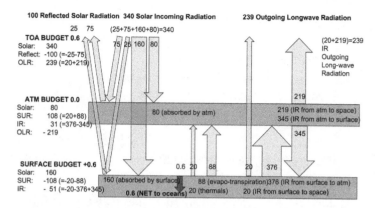

Figure 2.5. Energy fluxes of the incoming solar radiation, the infrared radiation emitted by the Earth's surface, and the energy fluxes from the Earth's surface to the atmosphere linked to mass movements (thermals) and evapo-transpiration processes. On the left-hand side, the three boxes illustrate the energy budgets at the top of the atmosphere (TOA), of the atmosphere, and at the surface. Note that the difference between the incoming solar radiation (340 W/m²), the reflected solar radiation (−100 W/m²), and the outgoing infrared, long-wave radiation is about +0.6 W/m², which is the amount of energy that the Earth (mainly the oceans) absorbs. (Data source: Hartmann 2006)

amount of energy that the Earth (mainly the oceans) absorbs. The three boxes on the left-hand side of Figure 2.5 summarize the energy budget at the top of the atmosphere (TOA), of the atmosphere and at the surface:

TOA (+0.6 W/m²):

- Solar incoming radiation: 340
- Reflected incoming solar: −100 (= −25 − 75)
- Outgoing long-wave IR radiation: 239 (= 20 + 219)

Atmosphere (0 W/m²):

- Solar incoming radiation: 80
- Surface (convection): 108 (= 20 + 88)
- Incoming long-wave IR radiation: 31 (= 376 − 345)
- Outgoing long-wave IR radiation: −219

Surface ($+0.6$ W/m^2):

- Solar incoming radiation: 160
- Surface (convection): -108 ($= -20 - 88$)
- Outgoing long-wave IR radiation: -396 ($= -20 - 376$)
- Incoming long-wave IR radiation: $+345$

2.7 What are the key similarities and differences of the atmosphere and the ocean?

The atmosphere and the ocean are both fluids that obey the laws of fluids. Three of their key differences are their density, the fact that the ocean's flow is constrained by physical boundaries, and the different relative roles that buoyancy, gravity, and the Coriolis force play in the two fluids.

These two fluids, the atmosphere and the global ocean, interact continuously and determine the weather and the climate. The oceans contribute to regulate the temperature of the lower layers of the atmosphere, and the atmospheric winds are in large part responsible for the circulation of the ocean.

The two fluids have a very different density: about 1,020–1,029 kg/m^3 for the ocean and about 1.2 kg/m^3 for the standard atmosphere (the atmosphere at 1,013 hPa and at 15°C). Note also that the atmosphere contains a tiny fraction of the total water of the Earth system: while 97% is contained in the oceans, only about 0.001% is contained in the atmosphere.

The distribution of land and its topography (i.e., the arrangement of the physical features of an area) affect both the lower layers of the atmosphere and of the oceans. In the oceans, the asymmetry of the land distribution between the two hemispheres and the sea floor topography determine the deep ocean currents, and how the ocean distributes heat from the warmer tropical regions to the colder northern and southern latitudes. In the atmosphere, the topography affects the lower layers of the atmosphere and can trigger vertical motions and convection events.

The oceans play a major role in determining weather phenomena and the climate because of their heat capacity and thermal inertia, and also because they cover about 70% of the Earth's surface. The atmosphere picks up moisture and heat from the oceans: ocean warm currents, such as the Gulf stream in the northern Atlantic Ocean, affect the atmospheric circulation and the regions where deep, cyclonic systems develop. In the tropics, cyclones and hurricanes develop where the ocean temperatures are higher, and heat fluxes from the ocean to the atmosphere can favor their development and sustain their propagation.

In the troposphere, the lower layers of the atmosphere, the flow circulation is determined and driven by heat transfer. Vertical motions, which cause convection, induce horizontal motions due to the Coriolis effect (linked to the fact that the Earth is a rotating sphere). Warm air rises above warm surfaces and descends when it is cooled. Generally speaking, warm masses of air rise in the troposphere over the tropical region, propagate northward toward the polar regions, cool down, and then descend toward the surface. The Coriolis effect induces motions along the parallels, and the combination of vertical and horizontal motions, and their interactions with topography, generates the small-scale features that constitute the daily weather. These features are location dependent, since the local topography and the surface characteristics (the presence of water bodies, of vegetation of different kinds, of cities, of land-sea contrasts, of mountains) affect them. Because of all these interactions and phenomena, different geographical areas even with the same latitude can experience rather different weather and climate conditions.

The Earth's rotation plays a very important role in determining the atmospheric and oceanic circulation. Consider, for example, the atmosphere: if the Earth was not rotating, its motions would be dominated by large-scale convection cells, with updrafts over the tropics and downdrafts over the northern and southern latitudes, and by much fewer motions along the parallels than with a rotating Earth. If the Earth had no oceans and a flat land surface, the climate of regions with the same latitude would be very similar, with fewer longitudinal variations than we observe today. Climatic zones would more or less run in belts parallel to the equator. Topography and the presence of the oceans and land-sea contrasts cause climatic zones to depend not only on latitude.

The ocean has two main types of circulations: a "wind-driven" circulation that affects its surface and a "thermohaline" circulation that affects its deeper layers. The surface circulation is strongly influenced by the atmospheric winds, which drag the surface of the oceans, and cause ocean waves and currents in its top layers, up to a depth of about 50–100 meters. Surface ocean currents are affected by the Coriolis effect and by the land-sea contrast. At middle latitudes (say, between 30–75°N and 30–75°S), the ocean surface currents run generally eastward, flowing clockwise in the Northern Hemisphere and counterclockwise in the Southern Hemisphere. Close to the polar region in the Southern Hemisphere, where the continents do not create any obstacle to the ocean circulation, ocean currents can run undisturbed from west to east.

Deep-water currents originate in a few regions of the polar oceans, where the creation of sea ice releases salt in the water, which causes the water to become sufficiently dense and heavier than the surrounding waters, to sink to the bottom floor. The polar regions where dense, cold water sinks to the bottom of the ocean are in the Atlantic Ocean, and in the Pacific Ocean close to Antarctica (in the Northern Hemisphere, the Pacific Ocean does not extend north enough). In the Northern Hemisphere, due to the conservation of mass, this sinking motion triggers a northward surface motion and an equatorward deep-water motion. Similarly, in the southern ocean, the sinking motion triggers a southward surface motion and an equatorward deep-water motion. Because of these motions, a vertical cross section of the average circulation of the whole Atlantic Ocean includes the following:

- A warm, wind-driven surface current moving from the equator toward the higher latitudes (in the northern Atlantic, it is called the Gulf current).
- The North Atlantic Deep-Water current, which originates east of Greenland, sinks to the bottom of the ocean and flows toward the equator.
- The Antarctic bottom water circulation, which originates along the ice edge of the Weddell Sea, sinks to the bottom of the ocean, and flows toward the equator.

In terms of velocities, there is a huge difference between the surface and the deep-water currents. The typical velocities in the wind-driven currents are about 0.1 m/s (i.e., about 0.36 km/h), with values reaching about 1 m/s (3.6 km/h) only in the narrow passages; the typical velocity of the deep-water currents is about 10–20 km a year (thus about 350 times slower).

Another important difference between the atmosphere and the ocean is linked to the relative role that buoyancy (i.e., vertical motions linked to difference in temperature and/or density of the masses of adjacent fluids), gravity, and the Coriolis force play in determining the fluid motions. The difference can be highlighted by a parameter that characterizes the dynamics of a fluid, the Rossby radius of deformation. This radius is the length scale at which rotational effects linked to the Coriolis force are as important as buoyancy or gravity. The Rossby radius of deformation is also the typical length scale of the large-scale phenomena of the two fluids, the length scale that a model built to study and predict the evolution of the two fluids needs to resolve to be able to realistically simulate the fluids' behavior.

In the atmosphere, the Rossby radius of the typical large-scale phenomena that occurs in the troposphere is about 1,000 km. For the oceans, the Rossby radius varies with the latitude, from about 100 km in the tropics to about 10 km at high latitudes. Thus, compared to the atmosphere, the Rossby radius is 10–100 times smaller in the ocean, which means that ocean models need to have a 10–100 times finer resolution than atmospheric models to be able to resolve in a realistic way the large-scale phenomena. This is the reason why the most accurate coupled ocean-atmosphere models used to study the climate have a higher resolution in the ocean than in the atmosphere.

2.8 Key points discussed in Chapter 2 "The Earth system"

These are the key points discussed in this chapter:

- In numerical weather prediction, an Earth system model is a model that includes all its components that affect the weather up to the forecast range we want to predict.
- The key building blocks of an Earth system model are the models of the atmosphere, of the global ocean, of the land, and of the cryosphere.
- Forecast accuracy and skill are assessed both by considering a few, very interesting case studies, selected to test the model in some particular weather situations, and by considering the average performance over many cases spanning at least one warm and one cold season.
- To continue to improve Earth system models and make them more realistic, we need to include all the relevant physical processes, where relevant means that they have a detectable impact on the model realism and on forecast accuracy and skill up to the forecast length we are interested in.
- Heat is transported and exchanged in the Earth system via radiation, conduction, and convection.

- The principal source of energy that drives the Earth system is the short-wave solar radiation: the Earth's temperature is determined by the balance between the incoming solar radiation and the outgoing infrared long-wave radiation.
- The atmosphere and the global ocean have many similarities and differences: weather phenomena and the Earth's climate are determined by their interaction.

3

OBSERVING
THE EARTH SYSTEM

In this chapter we discuss the role of observations, their key characteristics and inherent uncertainty, and how they are used to estimate the state of the Earth system. More specifically, we will be addressing the following questions:

1. Why do we need observations?
2. What are the key observation types?
3. Are observations affected by errors?
4. How do observation information and errors propagate?
5. Did COVID-19 affect weather forecast quality?
6. How do we observe the state of the atmosphere using satellites?
7. Do we have enough observations to determine the state of the Earth system?
8. Is it important to observe the whole atmosphere?

3.1 Why do we need observations?

We need observations to understand phenomena and validate our forecast models; we also need them to compute the initial state from where to start to generate a forecast.

Observing the Earth system is the first, key step in the numerical weather prediction process. Observations are essential to understand how phenomena evolve and to verify whether the models that we have built simulate reality in a sufficiently correct way. They are required to initialize weather forecasts, and to assess whether these latter are accurate and skillful. Advances in observations, mainly since the end of the 1970s, both in terms of quality and quantity, have been one of the key reasons why weather forecasts have progressed so remarkably. Observations are also essential for reconstructing the past climate. Observations can be classified accordingly to the variables that they relate to or to the characteristics of the instruments that take them.

3.2 What are the key observation types?

There are many observation types, differing in terms of the variables they refer to and the instruments that are used to take them. One way to classify them is according to the platform on which the instrument that takes them is mounted.

At the time of writing (2022), every day about 600 million weather observations are taken throughout the world and exchanged in real time, so that any national weather service can, if they wish and have the capabilities, monitor the on-going weather and generate a weather forecast. Most of these observations, let's say about 95%, are taken by instruments flying on satellites, while the rest rely on instruments onboard ships or located at ground stations, or are taken by balloon radiosondes.

Instruments can be grouped into three classes:

- *In situ instruments that take a measure at a specific point.* This class includes the conventional instruments (e.g., the mercury barometer, thermometers, or anemometers that measure the wind) and the radiosonde instruments that measure temperature, pressure, humidity, some chemical species, and aerosols.
- *Instruments that sample remotely an area of the Earth's surface, or a volume.* They measure the transmission

properties of the Earth's atmosphere at different wavelengths. Because the transmission properties depend on the air characteristics (e.g., temperature, water vapor and gases' concentrations, aerosols, sea surface temperature, sea waves), from these measurements we can deduce the air properties. These instruments can be ground based, aircraft based, or mounted on satellites. In an active observing system, electromagnetic pulses are transmitted through the atmosphere, and the reflected or scattered signal is then analyzed to deduce the atmosphere, or the Earth's surface, properties. In a passive observing system, the properties of the atmosphere are inferred from the radiation emitted, scattered, and/or reflected.

• *Instruments that calculate the wind by tracking objects moving in the atmospheric flow.* The object can be a radiosonde balloon tracked by a radar, a constant-level balloon tracked with a satellite navigation system, or clouds tracked by comparing consecutive images of geostationary satellites.

Each instrument has a different level of accuracy, which determines the observation error. For example, a good-quality mercury barometer used to measure surface pressure has an expected instrumental error of about 0.25 hPa (which corresponds to an error of less than 0.5% for average pressure values at sea level of about 1,000 hPa). For temperature, a surface thermometer has an error of about 0.1–0.2 K degrees (which corresponds to an error of about 0.05% for average temperature values of about 250–320 K degree), while the accuracy of a satellite temperature measurement is about 1 K degree.

A second type of observation error that needs to be considered is linked to the observation representativeness: this error depends on how an observation is used as part of an observation network, and it gives a measure of how accurately a single point observation can capture and describe the atmospheric flow. Suppose that we have a network of observations with an average spatial distance of 100 km, and we use it to observe convective precipitation (precipitation linked to convective systems). Since a convective system has a characteristic scale of about 1–10 km, if it happens between two stations, it will be missed by the observation network and will not be reported, thus leading to an underestimation of convective precipitation. If instead it occurs at the location of one station, we could think that convective systems are active on an area with a radius of about 100 km, thus leading to an overestimation of convective precipitation, while in reality it has occurred just at the station location. The representativeness error of a station observation provides an estimate of this source of uncertainty.

We can classify the observations depending on the platform on which the instruments are mounted and distinguish among the following:

- *Surface stations*: distributed irregularly over land and along ship routes, they usually observe surface pressure, temperature, wind components, and cloudiness, with high frequency (today often at least every 3 hours).
- *Radiosondes and pilot balloons*: they are launched by few surface stations usually every 12 hours; the variables that they observe are temperature, pressure, humidity, and wind components.
- *Aircraft reports*: they come from commercial aircrafts and usually include temperature and wind; they are spatially concentrated at airport locations and along the aircraft routes.
- *Vertical temperature soundings*: they are computed from radiance measurements taken by radiometers flying on polar orbiting satellites; by using different spectral channels, they can be used to deduce the temperature at different altitudes; these observations represent volume averages, with a horizontal size of tens of kilometers and a vertical span of a few hundred meters.

- *Cloud drift winds*: winds are derived by comparing subsequent images taken by geostationary satellites; algorithms are applied to identify cloud features and track them in time, so that the wind intensity and direction can be computed; each observation represents a volume-average wind.
- *Other observing systems*: other platforms that are used to take observations are drifting buoys, drop sondes, and constant-level balloons.

If we consider the remote sensing instruments, they are of two primary types, active and passive:

- Active sensors provide their own source of energy to illuminate the object they observe; the sensor emits radiation at a specific wavelength toward the object and then observes the radiation that is reflected or backscattered.
- Passive sensors detect the natural energy that is emitted, or reflected, by the object they observe; reflective sunlight is the most common source of radiation measured by the passive sensors.

The majority of active sensors operates in the microwave portion of the electromagnetic spectrum, which makes them able to penetrate the atmosphere under most weather conditions. Active instruments include the following:

- *Laser altimeters*: they are used to measure the height of the platform above the surface they overfly.
- *LIDAR (Light Detection and Ranging) instruments*: they emit a laser pulse, and by analyzing the reflected and/or scattered radiation, they can compute the distance to the object.
- *Radars*: they emit pulses of microwave radiation; by analyzing the backscattered radiation, it is possible to

reconstruct a two-dimensional image of the objects that reflected the radiation.

- *Scatterometers*: radars that emit high-frequency microwaves and are designed to measure the backscatter radiation; over the oceans, the backscatter radiation can be used to derive the wind speed and direction.
- *Sounders*: instruments that measure the vertical distribution of precipitation (and of any cloud meteor) and other atmospheric characteristics such as temperature, humidity, and cloud composition.

Passive sensors include different types of radiometers and spectrometers. Most of them operate in the visible, infrared, thermal infrared, and microwave portions of the electromagnetic spectrum. Passive instruments include the following:

- *Hyperspectral radiometers*: these sensors detect hundreds of very narrow spectral bands, so that they can discriminate among different targets.
- *Radiometers*: they measure the intensity of electromagnetic radiation in some bands.
- *Image radiometers*: they are radiometers with a scanning capability, so that their measurements can be used to build two-dimensional maps.
- *Sounders*: they measure the vertical distribution of atmospheric parameters, such as temperature, pressure, and composition.
- *Spectrometers*: they detect, measure, and analyze the spectral content of incident electromagnetic radiation.

3.3 Are observations affected by errors?

Yes, all observations are (and will always be) affected by errors and will never be perfect; these errors depend on the observation instrument. When used within a network to reconstruct the state of a system, they are also affected by the representativeness error.

Observations are affected by instrumental errors, linked to their construction and operation characteristics, and to the representativeness error, linked to how a network of observations is set and how they are used.

To understand how the observation errors affect the estimation of the state of the global atmosphere on a given day, suppose that we want to determine the surface temperature field on a regular grid, defined by one point every 10 km, and that we can rely on observations from two types of sensors:

- The first set of sensors is located at a few, specific stations and provides very accurate local observations.
- The second set of sensors is located on satellites, is less accurate, and provides observations that cover the area of the satellite swath.

Although the first set provides observations with very small instrumental errors, the fact that they are very few and sparse makes their representativeness error rather large when they are used to reconstruct the temperature field on the regular grid. The second set provides observations with a larger instrumental error, with no observations outside the satellite swath but a much better coverage over the satellite swath, with observations available even on a grid finer than the 10 km one, and thus a small representative error.

The Earth observing system in operation today provides about 600 million observations a day, with 95% of them from instruments onboard satellites. The 5% of observations taken from ground stations, buoys, ships, aircrafts, and radiosondes, although fewer in relative terms, are characterized by smaller instrumental errors than the satellite observations. The key problem that they have is that they are sparse and do not cover the globe in a uniform way: geographically, they are located mainly over land and along ship and aircraft routes. Aircraft observations are concentrated close to the airport locations, and radiosonde soundings are concentrated close to the launching sites. By contrast, satellite observations cover wide areas, but their instrumental error is larger, also because they could be affected by the sky conditions, and by the amount of water vapor, ice droplets, aerosols, and chemical components in the atmosphere. Since the concentration of these substances varies on a daily basis, their observation quality can also vary substantially every day.

So the 600 million observations give a very uneven coverage of the Earth system. The Northern Hemisphere is covered by many more land observations than the Southern Hemisphere: thus, estimating the state of the Earth system over the Southern Hemisphere relies mainly on satellite data. The quality of the estimation is also flow dependent, since it is more difficult over the areas affected by clouds and deep convection, due to the thick and vertically extended convection clouds. The overall effect of the observation network characteristics (instrumental and representative errors, coverage, observation sensitivity to the state of the atmosphere) is an uncertain estimation of the actual state of the atmosphere, with the uncertainty that varies between say 5% and 15% of the observed value; an uncertainty that is a function of the atmospheric flow.

3.4 How do observation information and errors propagate?

Observation information and errors propagate with the flow, at the same speed as weather phenomena propagate.

Observations contain essential information, necessary to estimate the state of the Earth system and to diagnose model performance. This is the reason why a lot of emphasis has always been put on observing the Earth system, maintaining the Earth observation system, and improving and upgrading it. Thanks to the development and launch of satellites, the number of Earth observations has increased by a factor of about 100 since the 1970s. At that time, the main sources of observations were land stations: this meant, for example, that observations of the atmosphere state over the oceans, which cover 70% of the Earth surface, were very limited in number. Furthermore, at that time very few observations of the vertical structure of the atmosphere were available. By contrast, today, thanks mainly to satellite observations, every day the three-dimensional structure of the global atmosphere is observed.

Numerical experiments have been performed to estimate how fast the information contained in a group of observations propagates. Suppose that all satellite observations over the North Atlantic Ocean are lost for a few days. As a consequence, the state of the atmosphere over the North Atlantic becomes very poorly known, since we can only rely on the very few observations taken by ships, aircrafts, and stations along the North Atlantic coast. Suppose that the rest of the world continues to be very well observed. Given that the average wind direction of the atmospheric flow is from the west to the east, we should expect that the errors linked to the loss of the satellite data will propagate mainly from the west toward the east.

How long will it take for errors over the Atlantic to propagate and affect the quality of weather forecasts over Europe?

Given that Europe borders the Atlantic Ocean, the impact is almost immediate. As the error of our estimate of the state of the atmosphere over the Atlantic deteriorates due to the missing satellite data, the forecast quality over Europe deteriorates.

Will the error propagate also further east, toward Asia? Will it impact the whole world?

Yes, the error will propagate further, but given that the rest of the world continues to be covered by good quality observations, the impact of the errors over the Atlantic will be felt less, the further away we move from the Atlantic.

This result was deduced from a comparison between two sets of weather forecasts: one starting from initial conditions computed using observations covering the whole world, and one with initial conditions computed with all but the satellite observations over the Atlantic Ocean. By comparing the error of the two forecasts, we monitored and tracked how the initial errors propagated from the North Atlantic region. Results showed that the error propagates eastward with a speed of about 2,000 km per day, and that the two forecasts differ substantially up to forecast day 2 to 3. After that time, the two forecasts would become rather similar: the reason is that after 3 days, the impact of initial errors elsewhere in the world and of model approximations dominate and mask the error due to the removal of the satellite observations over the Atlantic.

We can repeat this experiment considering the North Pacific Ocean instead of the North Atlantic Ocean. Let us remove all the satellite observations taken over the North Pacific Ocean for a few days and look at their impact on North America, and further to the east, on Europe. Results are similar for North America, in the sense that the propagation of initial errors over the Pacific Ocean dominates and affects the forecast quality over this region for up to about forecast day 3 to 5. They dominate for a longer time compared with the impact on Europe of removing observations in the Atlantic, because the North Pacific Ocean covers a much bigger area than the North Atlantic, and thus removing the satellite observations over this larger ocean has a more significant and longer-lasting effect. Because of the size of the Pacific Ocean, the impact on removing the satellite observations over the North Pacific can be detected also over Europe: the forecast started with all observations has a smaller error between forecast day 3 to 6 (which is the time it takes, roughly, for the error to propagate from the Pacific, over the United States, and across the Atlantic).

These observation simulation experiments are often performed to assess the impact, and estimate the value, of different types of observations. For example, one could remove all the observations taken from aircrafts, or the ones taken by one specific satellite instrument, or the observations taken by the same type of instruments onboard different satellites. These observation simulation experiments are repeated routinely because the value of the observations is intrinsically linked to the forecast model and the data assimilation used to estimate the initial conditions: for example, the same type of experiments, performed for the same period but with a poorer model and a different data assimilation system, would have given a different result.

3.5 Did COVID-19 affect weather forecast quality?

Yes, the fact that the number of good quality observations taken by instruments onboard commercial planes decreased substantially during COVID-19 had a measurable effect on the quality of weather forecasts.

A very interesting "real-time experiment" happened in 2020. Due to the coronavirus pandemic, 2020 saw a dramatic decrease in the number of flights, from a peak reduction of up to 80%, to an average reduction of about 60%. This had a major impact on the observations that were taken by instruments onboard these flights. Although we are talking of a maximum reduction of about 1% of the total number of observations, at the specific locations where the aircraft data are concentrated (cities with large commercial airports, busiest aircraft routes), the reduction was substantial, and the net impact was a dramatic increase in the local initial condition error, of up to 10%–20%. The comparison between forecasts issued before and after this decrease in the number of aircraft observations indicated an impact on forecasts up to day 3, with regional forecast errors increased by about 3%.

Is a 3% increase in error a large number or a small number?

To answer this question, we should consider that, on average, top-quality meteorological centers manage to reduce the average error of their short (day 1 to 3) and medium-range (day 3 to 15) forecasts by about 0.5%–1% every year. This reduction is achieved thanks to a combination of model development, better use of the existing observations, inclusion of new observations, and use of more powerful computers. Thus, 3% compared to say a 0.5%–1% average annual error reduction indicates that the quality loss was equivalent to 3–6 years of work.

3.6 How do we observe the state of the atmosphere using satellites?

Satellites observe the Earth using active or passive instruments, and they measure how radiance propagation is affected by the atmosphere.

Since the first meteorological satellite, TIROS (Television InfraRed Observation Satellite), was launched in 1960, satellite images have been used to give a picture of the weather from space, and they have become a familiar sight on television weather forecasts. Movie loops of images taken from geostationary satellites give a graphic impression of the movement of clouds. Viewed in the visible part of the spectrum, as the human eye would see from space, the shadows cast by the sunlight provide a three-dimensional view of the clouds, but only where there is daylight. During the night, infrared imageries provide a similar view of the cloud systems.

Today, satellite observations make a crucial contribution to the quality of weather forecasts (Figure 3.1). Their global coverage means that they provide information about the state of atmosphere, the land surface, the ocean, and sea ice that cannot be provided by the in situ measurements. When merged with modelled states through a process called data assimilation, they help to produce the best possible estimate of the current state of the Earth system. This estimate, called the analysis, is used as the initial conditions on which weather forecasts are based.

Figure 3.1 shows the relative number of observations from a range of platforms received by the ECMWF on March 1, 2017 (the percentages today are not very dissimilar from these):

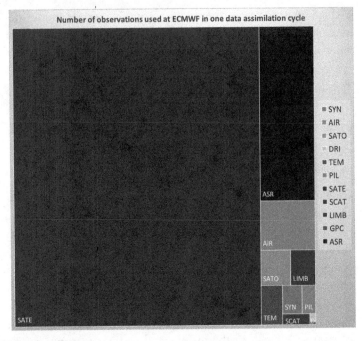

Figure 3.1. Relative number (percentage) of observations received daily at ECMWF (see text for a detailed explanation of the acronyms). (Source: ECMWF statistics based on data received on March 1, 2017)

- SATE: SATEM satellite sounding data: ~1,882M (81% of the total)
- ASR: satellite all-sky radiances ~243M (10%)
- AIR (AIREP): aircraft data ~69M (3%)
- SATO (SATOB): satellite atmospheric motion vectors ~27M (1.2%)
- LIMB: satellite limb observations ~22M (1%)
- TEM (TEMP): radiosondes ~21M (0.95%)
- SYN (SYNOP): land station and ships ~14M (0.6%)
- PIL (PILOT): balloons and profilers ~10M (0.4%)
- SCA: SCATT satellite scatterometer data ~7.3M (0.3%)
- DRI (DRIBU): drifting buoys ~1.1M (0.05%)
- GPC: ground-based precipitation composites ~0.4M (0.02%)

We can classify weather satellites in three categories:

- Polar orbiting satellites, which orbit the Earth close to the surface, and take six or seven detailed images a day
- Geostationary satellites, which stay over the same location on Earth high above the surface taking images of the entire Earth as frequently as every 30 seconds
- Deep space satellites, which face the sun to monitor powerful solar storms and space weather

Polar orbiting satellites fly at a typical height of 850 km (Figure 3.2), allowing their instruments to make measurements at far greater resolution than geostationary satellites. Orbiting the Earth about every 100 minutes, they scan wide swathes of the atmosphere as they sweep from pole to pole. Over the course of the day, they view most parts of the Earth at least twice. Their orbits are sun-synchronous, which means that they see that same part of the Earth at the same local time each day.

Geostationary satellites, located over the equator at a height of about 35,800 km, orbit the Earth once every 24 hours (Figure 3.3). Spinning at the same rate as the Earth, they stay above the same spot all the time and provide an unbroken series of images of the atmosphere below.

Satellites carry passive instruments, which measure radiation emitted naturally, and active instruments, which send out signals and measure the backscatter radiation. Natural radiation contains information on, for instance, temperature, humidity, clouds, and surface conditions. It can also provide information on winds by tracing motions of humidity or cloud features in successive observations. Active instruments use radar or lidar to probe the surface, clouds, and winds. Radio occultation observations are unique in that they involve sending signals from one satellite to another. The bending angle of such signals crossing the troposphere or stratosphere depends on temperature and humidity.

ECMWF data coverage (all observations) - IASI
2022071121 to 2022071203
Total number of obs = 156213

METOP-B (84223) ◆ METOP-C (71990)

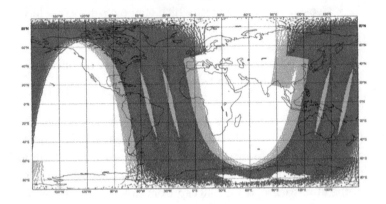

ECMWF data coverage (used observations) - IASI
2022071121 to 2022071203
Total number of obs = 22392

METOP-B (12070) ◆ METOP-C (10322)

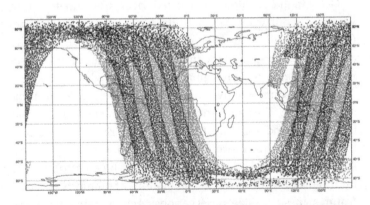

Figure 3.2. Infrared atmospheric sounding interferometer (IASI) observations taken by the European Union METOP-B and METOP-C satellites in the 6 hours between 21 GMT on July 11 and 03 GMT on July 12, 2022. The top panel shows all the data received at ECMWF, and the bottom panel shows the data used to generate the analysis valid for 00 GMT on July 12. (Source: ECMWF)

ECMWF data coverage (all observations) - GEOSTATIONARY RADIANCES
2022071121 to 2022071203
Total number of obs = 2094324

⬤ METEOSAT-9 (238438)   ◆ HIMAWARI-8 (346951)   ▲ METEOSAT-11 (242068)   ▼ GOES-16 (603302)
✕ GOES-17 (663565)

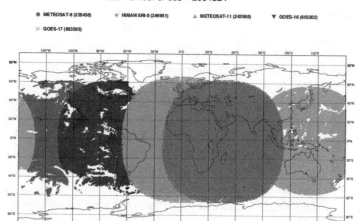

ECMWF data coverage (used observations) - GEOSTATIONARY RADIANCES
2022071121 to 2022071203
Total number of obs = 162735

⬤ METEOSAT-9 (21713)   ◆ HIMAWARI-8 (29110)   ▲ METEOSAT-11 (26272)   ▼ GOES-16 (38796)
✕ GOES-17 (46854)

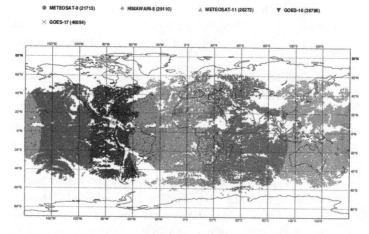

Figure 3.3. Geostationary radiance observations taken by five satellites in the 6 hours between 21 GMT on July 11 and 03 GMT on July 12, 2022. The top panel shows all the data received at ECMWF, and the bottom panel shows the data used to generate the analysis valid for 00 GMT on July 12. (Source: ECMWF)

The wide range of satellite observations collected to monitor the Earth and estimate the state of the atmosphere can be classified accordingly to the observation instrument:

- *Images*—The imagery from both geostationary and polar orbiting satellites are processed and analyzed to provide a wealth of information about the atmosphere and the surface of the Earth. For example, infrared imagery is used to measure the coverage and temperature of clouds and identify those that are rain-bearing. The speed and direction of the wind can be inferred from the apparent motion of clouds between successive images. The temperature of the sea surface can be measured from satellites, as can the extent of sea ice near the poles and snow cover over land. Fog, dust storms, pollution, and volcanic ash are other features that can be detected from space. Clearly, all of this is possible only by processing the satellite observations with models capable of translating the observed variables (radiation with different wavelengths) into variables that describe the state of the atmosphere, such as temperature, humidity, or wind.
- *Radiometers*—High-precision instruments on satellites measure the radiation emitted by the atmosphere and the Earth's surface. This is radiation naturally emitted by all bodies, typically in the infrared and microwave parts of the spectrum, which is related to the temperature of the body. Radiometers make accurate measurements of temperature and water vapor. Where the instrument is sensitive to a selected frequency in the infrared, the measurement is related to the average temperature of a cloud-free layer of the atmosphere. In the microwave part of the spectrum, the measurement is related to the temperature and humidity of the layer. An example of this instrument is the Infrared Atmospheric Sounding Interferometer (IASI) onboard the European Union METOP-B and METOP-C satellites. Microwave measurements, unlike

their counterparts in the infrared, provide valuable information in the presence of clouds. By having a number of radiometers on the satellite, each sensitive to a different frequency, it is possible to derive the vertical profiles of temperature and humidity through much of the atmosphere. These vertical profiles are called soundings. Having global coverage, they are of immense value to numerical models that forecast the weather.

- *Scatterometers*—A scatterometer is a radar system mounted on a polar orbiting satellite that directs radar pulses toward the Earth's surface and measures the strength of the backscattered return beam. Over the oceans backscattering is caused by small wind-generated waves. By measuring the backscatter at two or more angles of incidence, it is possible to derive the wind speed and direction close to the sea surface.

- *Global Positioning System (GPS) observations*—The signals from global navigation satellites that help us navigate our way in a car have meteorological applications as well. Since the time of arrival at a ground station of a satellite signal passing through completely dry air can be calculated with immense accuracy, and since water vapor in the atmosphere slows its arrival, by measuring the delay we can estimate the total water vapor content along the signal's path from the satellite. These data are particularly valuable for identifying areas of deep moist air linked to thunderstorms on hot summer days. As well as being received by a ground station on Earth, the signal from a global navigation satellite may also be received by a second low-orbit satellite providing a different source of meteorological data. As the satellite is occluded by the passage of the Earth's atmosphere, the signal's path is bent by atmospheric density gradients. The variation in the bending with the height above the surface provides valuable information about the temperature and humidity through the depth of the atmosphere.

- *Altimeters*—Some satellites in low polar orbits carry an altimeter that can measure very accurately the distance between the satellite and the Earth's surface below its path. These measurements have several valuable applications, particularly over the oceans. The height of the ocean surface is dependent on many factors such as the physical properties of the ocean (temperature, salinity, waves), oceanic tides, atmospheric pressure, and local gravitational anomalies of the Earth's crust. Areas of the ocean where the surface height is uniformly higher than average may be attributed to warm columns of water which expand as their temperature rises. Currents on at the ocean surface are driven by gradients in the height of the ocean surface in the same way that winds in the atmosphere are driven by gradients in surface pressure. The surface current may therefore be inferred from altimeter measurements in the same way that the wind may be inferred from measurements of pressure. Other meteorological and oceanographic applications of altimeter data include measurement of wave height and wind speed at the ocean surface, the thickness and extent of sea ice, and the rate of rise in global sea level over the long term.
- *Doppler wind lidars*—A lidar emits pulses of light from a laser and measures the properties of targets from the nature of the returned signal. In the case of a Doppler wind lidar, the targets are atmospheric molecules, cloud droplets, and aerosols. The measured Doppler shift is directly related to the wind speed along the line of sight of the lidar beam. If two or more lidar beams are used, pointing in different directions, both wind speed and direction can be obtained.

Today, the vast majority of satellite data used to compute the state of the atmosphere and initialize weather forecasts comes from passive instruments, measuring infrared or microwave radiances. These radiances typically reflect conditions in a rather deep layer of the atmosphere and could be the result of any number of atmospheric states. To be able to compare the observations with the state of the atmosphere in the forecasting model, we need to know the "simulated radiances," that is, the radiances that would be observed if the model state correctly described the atmosphere. These "simulated radiances" are computed using a numerical model component called the "observation operator" that generates the radiances that the satellite would measure if it observed the "digital" atmosphere simulated by the model.

By comparing the simulated radiances with the observed ones, data assimilation computes how the model state needs to be adjusted to bring it closer to the observations. Data assimilation is performed over a time window of a few hours: all observations collected in this window are used, and they are compared to a short-range weather forecast that spans this window. The use in operational weather prediction of accurate data assimilation procedures is key to be able to reduce the initial condition errors to the minimum value, and thus start a weather forecast from a state that is very close (say, as close as possible given the collected observations and the modelling and assimilating capabilities) to the real state of the atmosphere. Failure to do so would introduce initial error that could rapidly amplify and propagate, and thus affect forecast quality.

3.7 Do we have enough observations to determine the state of the Earth system?

No. Some parts of the Earth are still covered by a limited number of observations or by low-quality observations: the oceans, in particular, are still very poorly observed.

The past 40 years have seen a very rapid increase in the number and quality of Earth observations. This increase has been primarily due to satellite observations that today constitute about 95% of all the data that are collected and exchanged daily to generate weather forecasts. Thus, while before the satellite era there were parts of the atmosphere with very few observations (e.g., the atmosphere above the oceans, especially over the polar regions), today we have a more homogenous observation coverage of the atmosphere. Today, about 600 million observations are taken and shared every day, and 95% of them are from satellites. Still, there are parts of the globe that are less well observed, or where observations have a lower quality, and some components of the Earth system that are less observed than others.

If we consider the deep ocean, for example, every day we collect only a few tens of thousands of observations, compared to the several millions of observations taken in the atmosphere. Because of this, the knowledge of the deep state of the ocean is less accurate than the knowledge of the atmosphere state.

Considering the atmosphere, land-based observations, observations taken by ships (Figure 3.5) or aircrafts (Figure 3.6), are of a higher quality than satellite observations, but they provide good-quality observations only over a few, specific locations. Satellites cover larger areas, but due to the configuration of the satellite orbits, their observations do not have the same quality everywhere: for example, the polar regions are covered by less accurate observations. Furthermore, the geostationary satellites are positioned to observe the weather only mainly in the mid-latitudes (Figures 3.2, 3.3, and 3.4) and do not cover the polar regions. Thus, overall, there are still parts of the Earth that are not observed as well as others.

It is also worth mentioning that only about 15% of the observations collected today are used to estimate the state of the atmosphere, because assimilating all of them would require much more computer power than we have available today. This is especially true for the satellite data, as you can see by comparing in Figures 3.2–3.6, the top "a" panels, which show all received data at ECMWF, against the bottom "b" panels, which show the used data. Future computer power increases should allow the use of a larger percentage of the observations.

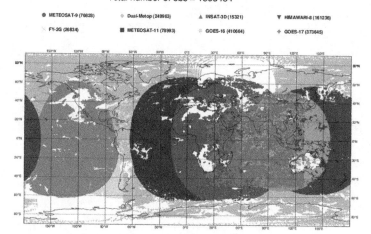

ECMWF data coverage (all observations) - AMV IR
2022071121 to 2022071203
Total number of obs = 1393484

● METEOSAT-9 (76628)   ⬦ Dual-Metop (249963)   ▲ INSAT-3D (15321)   ▼ HIMAWARI-8 (161236)

FY-2G (26834)   ■ METEOSAT-11 (78993)   ⬦ GOES-16 (410664)   ⬦ GOES-17 (373645)

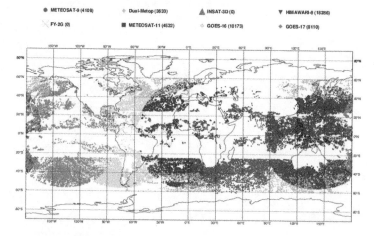

ECMWF data coverage (used observations) - AMV IR
2022071121 to 2022071203
Total number of obs = 48832

● METEOSAT-9 (4108)   ⬦ Dual-Metop (3633)   ▲ INSAT-3D (0)   ▼ HIMAWARI-8 (18286)

FY-2G (0)   ■ METEOSAT-11 (4522)   ⬦ GOES-16 (10173)   ⬦ GOES-17 (8110)

Figure 3.4. Atmospheric motion vector (AMV) observations taken by eight satellites in the 6 hours between 21 GMT on July 11 and 03 GMT on July 12, 2022. The top panel shows all the data received at ECMWF, and the bottom panel shows the data used to generate the analysis valid for 00 GMT on July 12. (Source: ECMWF)

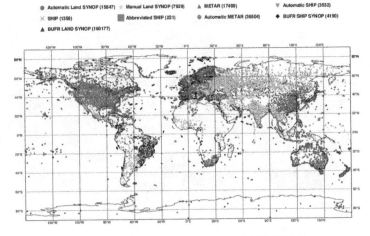

ECMWF data coverage (all observations) - SYNOP-SHIP-METAR
2022071121 to 2022071203
Total number of obs = 247245

Automatic Land SYNOP (15847)   Manual Land SYNOP (7929)   METAR (17489)   Automatic SHIP (3532)
SHIP (1356)   Abbreviated SHIP (221)   Automatic METAR (36504)   BUFR SHIP SYNOP (4190)
BUFR LAND SYNOP (160177)

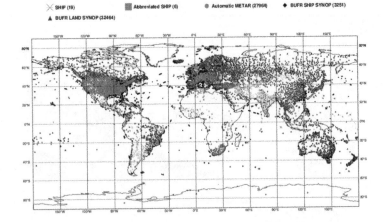

ECMWF data coverage (used observations) - SYNOP-SHIP-METAR
2022071121 to 2022071203
Total number of obs = 85864

Automatic Land SYNOP (2336)   Manual Land SYNOP (2333)   METAR (16696)   Automatic SHIP (623)
SHIP (16)   Abbreviated SHIP (6)   Automatic METAR (27964)   BUFR SHIP SYNOP (3251)
BUFR LAND SYNOP (32464)

Figure 3.5. SYNOP observations taken in the 6 hours between 21 GMT on July 11 and 03 GMT on July 12, 2022. The top panel shows all the data received at ECMWF, and the bottom panel shows the data used to generate the analysis valid for 00 GMT on July 12. (Source: ECMWF)

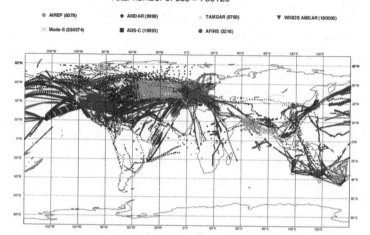

ECMWF data coverage (all observations) - AIRCRAFT
2022071121 to 2022071203
Total number of obs = 780125

AIREP (6078)   AMDAR (9998)   TAMDAR (5760)   WIGOS AMDAR (180608)
Mode-S (554574)   ADS-C (19693)   AFIRS (3216)

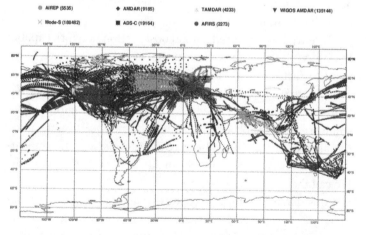

ECMWF data coverage (used observations) - AIRCRAFT
2022071121 to 2022071203
Total number of obs = 364016

AIREP (5535)   AMDAR (9185)   TAMDAR (4233)   WIGOS AMDAR (135144)
Mode-S (188482)   ADS-C (19164)   AFIRS (2273)

Figure 3.6. Aircraft observations taken in the 6 hours between 21 GMT on July 11 and 03 GMT on July 12, 2022. The top panel shows all the data received at ECMWF, and the bottom panel shows the data used to generate the analysis valid for 00 GMT on July 12. (Source: ECMWF)

3.8 Is it important to observe the whole atmosphere?

We need to have a good coverage of the whole atmosphere, although it is true that there are regions of the atmosphere that are more important to be observed than others.

It is clearly important to have a good coverage of the whole atmosphere: indeed, one of the reasons why weather forecasts over the Southern Hemisphere were less accurate than the forecasts over the Northern Hemisphere up to about 20 years ago is that there was a huge difference between the number and the quality of the observations covering the two hemispheres. Once satellite data started covering the Southern Hemisphere well enough, the difference became less evident, and today the quality of forecasts of the same variable over the two hemispheres is very similar. Not identical, since the two hemispheres have different characteristics (e.g., in the percentage of the hemisphere covered by land and sea, or in the position of the mountain ranges), and thus different types of dominant circulation patterns, but they are much closer than in the past (Figure 3.7).

Once we have ensured a good observation network capable of observing the large-scale features of the atmosphere, if we want to further increase the observation coverage, we could consider that some regions are more important to be observed than others for weather prediction. These are the regions where small errors in the observations, and thus in the estimate of the initial state of the atmosphere, could grow faster and thus affect the forecast quality more quickly and more dramatically. We call them "sensitive regions."

Consider, for example, a weather map of the Atlantic Ocean in winter, with a low-pressure system developing over the ocean and propagating toward western Europe. If we want to predict its evolution toward Europe, we have to observe very accurately its state, and the state of the key meteorological features that could affect its evolution, for example, the position and intensity of the mid-latitude jet stream. An

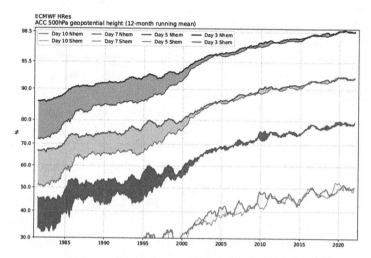

Figure 3.7. Quality of the ECMWF forecasts over the Northern Hemisphere (top, bold lines) and over the Southern Hemisphere (bottom, thin lines). Curves with different shading of gray indicate the quality of forecasts of the mid-level atmospheric flow, represented by the geopotential height at 500 hPa and measured by the anomaly correlation coefficient (ACC ranges from 0 to 1, with 1 for a perfect forecast). The top curves refer to the day 3 forecasts, the second level of curves refers to the day 5 forecasts, the third level of curves refers to the day 7 forecasts, and the bottom curves refer to the day 10 forecasts. For each year, each curve indicates the anomaly correlation coefficient (ACC) of the 500 hPa geopotential height forecast: for example, in 2015, the day 7 forecasts valid over both hemispheres had an average ACC of about 0.75. (Source: ECMWF)

intensifying low-pressure system evolves very rapidly, and small errors in estimating its state (position and shape, intensity, humidity content, associated wind) could lead to very large forecast errors in a very short time. By contrast, having little or inaccurate observations in other parts of the Atlantic Ocean that do not affect the propagation of this low-pressure system would have a very little impact on our forecast. Using sophisticated modelling techniques, we can identify the "sensitive regions" and thus aim to increase the number and quality of observations in these regions.

This example illustrates that not all regions of the atmosphere need to be observed with the same accuracy. Indeed, often extra observations are taken only in what are called the "atmospheric flow sensitive regions." This happens every year when tropical cyclones develop in the tropical western Atlantic, propagating toward the Gulf of Mexico and the U.S. East Coast. To improve our knowledge of these cyclones, and thus generate more accurate forecasts, extra observations are taken every few hours. Similar studies of the impact of targeted observations on forecast quality were conducted in the past over the northern Atlantic: "most sensitive areas" had been identified where to take extra observations, to improve the prediction of extra-tropical cyclones and windstorms hitting Europe. Results demonstrated that adding these extra, targeted observations could reduce the error of a 1-to-3-day forecast by up to about 10%.

The fact that it is more important to observe some regions of the atmosphere than others means that, if we have a limited amount of resources, we should concentrate them to observe those "sensitive areas" where initial errors, if present, could grow faster. Considering satellite data, given that we can only assimilate a certain number of observations (because we do not have enough computer power to assimilate all of them), we could plan to use and assimilate more observations located in the sensitive regions, and less outside them. In other words, we could operate with a "flow-dependent" observing system: even if the collected observations do not change daily, the ones that are selected to be assimilated could depend on where they are localized.

3.9 Key points discussed in Chapter 3 "Observing the Earth system"

These are the key points discussed in this chapter:

- Weather observations can be grouped depending on the instruments used to take them in three classes: in situ instruments, instruments that sample an area of the Earth surface or a volume remotely, or instruments that calculate the wind by tracking objects moving in the atmospheric flow.

- Observations also can be classified according to the platforms onto which the instruments that take them are mounted.
- Observations are affected by instrumental and representativeness errors, and they will never be perfect.
- Satellites differ in their orbit and in the type of observations they collect.
- Today, about 600 million observations are taken and shared in real time every day; 95% of them are from satellites; about 15% of these observations are used daily to estimate the state of the Earth system.
- Observations are essential to estimate the state of the atmosphere and to design, diagnose, and improve Earth system models.
- Since neither the quality nor the area covered by the Earth observations is uniform, each region of the globe is characterized by different observation errors: the net effect is that the error of the estimate of the state of the Earth system is not uniform.
- There are some "sensitive regions" of the atmosphere that are more important than others: observations should cover well these regions to provide a very accurate initial state for weather forecasts.

4

MODELING
THE EARTH SYSTEM

In this chapter we explain what a model of the Earth system is, what its key components are, and on which equations it is based. We also discuss what it means to solve the equations numerically. More specifically, we will be addressing the following questions:

1. From where should we start to model the Earth system?
2. What are the state variables of a system?
3. How many variables define the state of the Earth system?
4. What is an equation?
5. What are analytical and numerical solutions of an equation?
6. How do we deduce the equations used to study the weather and the climate?
7. What is an example of a set of equations that can predict the weather?
8. How do we solve the primitive equations?
9. What is the difference between a prognostic and a diagnostic equation?
10. Can we use simple models to understand the behavior of complex systems?

4.1 From where should we start to model the Earth system?

Physics is where we must start to build a model of the Earth system capable of simulating realistically its key phenomena.

By applying the principles of conservation of mass, momentum (i.e., the product between the mass of a fluid parcel and its velocity), and energy, we can derive a set of equations that describes the flow of a fluid such as the atmosphere or the ocean. On top of the equations deduced by applying the conservation laws, an Earth system model also includes relationships between the variables that define the state of the atmosphere, such as the law of gases, which relate the density, pressure, and temperature of the atmosphere. A model also includes equations that describe processes such as vegetation, or the water cycle, or how orography can induce turbulent motions. All these equations are deduced by applying the laws of physics and chemistry.

Mathematics (e.g., numerical methods that can be applied to solve numerically an equation) and software technology (e.g., to write efficient codes that can exploit the architecture of the modern computers) are also applied to develop good-quality and efficient numerical models of the Earth system.

4.2 What are the state variables of a system?

With the term "state variables" we mean the variables that define "completely" the state of a system: for a fluid, for example, we are talking about temperature, wind, pressure, density, and humidity at key, relevant locations.

The type and number of the state variables that describe a physical system depend on the system itself: simple systems could be described by a very small number of variables at a few key locations, while complex systems, such as the Earth system, require the knowledge of many more variables at a very large number of locations. The complexity of a system has an impact on the number of the state variables.

Consider, for example, a ball moving on a billiard plane (let us assume the ball cannot "jump" outside this plane but can only move horizontally, and that we know its mass). There are four variables that define its state: its position and velocity measured with respect to a coordinate system defined, for example, by two perpendicular sides of the billiard. If we know these four variables, we can define the state of the system (the ball): we can observe it and describe how it changes its state. Suppose that we want to describe how the ball moves when it is subjected to a force, for example, if we hit it with a cue. From the force imparted by the cue we can compute the ball's acceleration, and from the acceleration and the initial velocity we can compute how its velocity varies with time. If we know the ball's initial position and its initial velocity, we can then compute how it is going to move. Thus, from the initial conditions, the ball's mass and the force imparted by the cue, we can predict how the ball will move on the billiard.

Let us now consider the atmosphere, a more complex system. We can describe the state of a moist atmosphere at a single point of a three-dimensional mesh that covers the whole globe with the following seven variables: temperature, air density (i.e., mass divided by volume), the three wind (i.e., velocity) components, the water vapor content, and surface pressure.

Suppose that we want to compute how the atmosphere in a region around where we live, say, a region of about 500 × 500 km, will evolve in time. In other words, we want to make a weather forecast for this region. To be able to do so, we need to know the initial state of the atmosphere on a much bigger area than the one we are interested in, since air masses propagate in time, and the weather inside this region in the future will be affected by the current weather outside the region.

Air masses propagate about 2,000 km a day: thus, if we want to predict the weather where we stand for the next 3 days, we need to know as accurately as possible the state of the atmosphere within a large area that extends at least for about 6,000 km to the west, since the air that 3 days ago was 6,000 km from us, during that time propagates and influences the local weather we are having. Given that the atmospheric flow has also a meridional direction (i.e., a component along the North-South direction), the area needs to extend also in the meridional direction, but since the average meridional velocity is smaller than the average longitudinal velocity, the area extension along the meridional direction can be smaller than the extension in the longitudinal direction. Thus, the area that we need to consider is more like a rectangle elongated along the longitudes, rather than a circle centered where we stand. Note also that we need to extend the integration domain to the east, to limit the error backward propagation from the eastern border. Thus, let us say that we need to consider a domain that is about 8,000 km × 6,000 km. We also need to consider that an air mass changes position in the vertical dimension: this implies that we need to know its state not only on a two-dimensional plane where of interest, as it was the case for the billiard ball, but inside a whole three-dimensional volume.

Once we know the initial state of the atmosphere inside this large volume, we can predict how it evolves in the future at our point of interest for the whole forecast length (3 days in this example). This large volume becomes even larger if we want to predict the weather for a longer time.

How many grid points do we need to simulate the weather? And thus how many variables do we need to know to describe the state of the system?

Suppose that we represent the state of the system inside this volume on a finite mesh with one grid point every 50 km and over 30 vertical levels: this means that the 8,000 × 6,000 km area of integration includes 576,000 points (160 × 120 in the horizontal, × 30 levels in the vertical): if we multiply this number by the number of variables that describe the state of the system at a single location, we see that the state of the system is described by about 4 million state variables.

4.3 How many variables define the state of the Earth system?

If one considers that for the atmosphere all scales matter, then the number of state variables is infinite. But since we can only describe the state of the Earth system with a finite number of variables, that number depends on the complexity and resolution of the Earth system model. In numerical weather prediction, the state-of-the-art global models typically accomplish this on a finite three-dimensional grid with a grid spacing of about 10 km and about 150 vertical levels, which means with a number of state variables on the order of ten billion (10^{10}).

We have discussed earlier that the state of the simple system "two-dimensional billiard ball" is described by four variables: its two-dimensional position and velocity vectors. Once we know its initial position, velocity and acceleration, the force that is applied to the ball by a cue, and the mass of the ball, we can predict how and where it will move.

We have seen earlier that the state of the atmosphere on three-dimensional mesh that covers a 8,000 × 6,000 km area with one point every 50 km, and that extends up to 20 km height with 30 levels in the vertical, has 576,000 grid points (160 × 120 × 30). At each point, we need to know the air density, temperature, water vapor, the three velocity (wind) components, and the surface pressure. Thus, the number of variables is about 4 million.

Now suppose that we want to predict the state of the atmosphere for up to 15 days, on a mesh with a horizontal grid spacing of 10 km, and with 140 vertical levels that extend to a height of up to about 80 km. Since the forecast length is longer than a few days, we cannot limit ourselves to a finite area but need to use a global model. A mesh that covers the whole globe with a resolution of 10 km has about 4,000 points on a longitude parallel at the equator (calculated as the circumference of the Earth at the equator in kilometers divided by 10,

$$n = \frac{2\pi r_e}{d} = \frac{2 \cdot 3.14 \cdot 6,370}{10}$$. which gives 4,000) and has 2,000

parallels. Thus, this mesh has about 4,000 × 2,000 × 140 grid points, which means about 1.12 billion grid points: actually, since as we move closer to the poles, the length of a parallel diminishes, the number of points of this mesh is about 25% less, thus let us say that the mesh has about 1 billion points. If we multiply this number by the number of variables, we see that the number this system is defined by has about 7 billion variables, that is, 7,000,000,000.

In fact, 10 billion is approximately the number of state variables of the state-of-the-art global models used at the most advanced meteorological centers, since they include also other variables, which are introduced to define more accurately the state of the atmosphere (e.g., the concentration of three types of cloud meteors: water droplets, snow flaxes, and ice crystals).

4.4 What is an equation?

An equation is a relationship between variables.

Some equations describe a relationship between variables at a specific time: for example, they could describe how to compute the pressure of a gas if you knew its density and temperature, or they could compute the flux of moisture or of heat from the deep soil to the atmosphere due to the presence of different types of vegetation. Other equations describe how variables change in time when they are subjected to variations of energy or of momentum (e.g., to forces). The Newton law of motion is of this second kind, and it can be used to compute the motion of a particle, or of a mass of air, under the influence of forces.

If we apply the laws of physics to the atmosphere, we can deduce a set of equations that describe how the state variables evolve at each grid point of the three-dimensional mesh that we use to describe the system.

There is not a unique set of equations that describe the Earth system: they depend on the scales that we want to predict, on the truncation we apply, on the observations that we can access to estimate the initial state of the system, on the methods used to solve them, and, if the equations are solved numerically, on the computing resources available. They also depend on the approximations that we decide to apply to the original equations when they are transformed into a numerical code to be solved.

In fact, starting from the same laws of physics and principles, one can deduce a hierarchy of models, all capable of describing the behavior of the atmosphere with a different degree of realism. The more the equations are simplified, the less realistic the model is, the more limited the number of scales that it can describe is, and, in general, the less accurate the predictions are.

Sometimes approximations have to be applied because it is impossible to initialize all the degrees of freedom of the system; that is, it is impossible to estimate in an accurate way the state variables at all the grid points of the three-dimensional mesh onto which the equations are solved. At other times, approximations have to be made because it is impossible to verify whether all the terms of the equations that have been kept are correct, for example, because of the lack of observations to assess this. Approximations could also be required because it is impossible to solve the full set of equations numerically because of the lack of computing resources.

4.5 What are analytical and numerical solutions of an equation?

An analytical solution of an equation is a solution that can be written as a combination of mathematical functions, while a numerical solution is one expressed by a number, or a set of numbers, computed by a (super) computer.

The simplest equations can be solved analytically, which means that one can write formulas that involve sums and products of mathematical functions that satisfy the equations. Analytical solutions have been found only for very simple, low-dimensional sets of equations, but not for the complex ones that describe the behavior of systems even with just a few tens of degrees of freedom. The only way to solve the equations that describe high-dimensional and complex systems is numerically, using a very powerful computer that can make the billions of calculations required to generate a forecast in a reasonable amount of time.

One very important aspect to consider is that a computer calculation is affected by "round-off errors"; that is, its accuracy is dictated by how powerful its central processing unit (CPU) is. Thus, even if we could write an exact numerical version of a set of equations, the numerical solution would never be identical to the analytical solution. The impact of the CPU characteristics is negligible if one solves simple stable systems, whereby the numerical solution can be computed with a small number of operations and small errors do not grow fast. But the impact can be substantial, and lead to large errors. If the system is complex, billions of operations have to be computed, and small errors can grow quickly.

On top of round-off errors, often approximations are made when transforming an equation into lines of computer code. The combination of the round-off errors and of the numerical approximations leads to model errors that affect a forecast.

To give a very simple example, suppose we want to compute the product between two numbers, $x = \dfrac{3}{7}$ and $y = \dfrac{14}{24}$.

We can compute the product analytically:

$$x \cdot y = \frac{3}{7} \cdot \frac{14}{24} = \frac{3 \cdot 2}{24} = \frac{1}{4} = 0.25.$$

Alternatively, we can first compute a and b numerically, and then multiply them. Suppose that we do this with a computer that can keep in memory only the first six digits after the unit. This computer will give:

$$x_m = 3/7 = 0.428571.$$

$$y_m = \frac{14}{24} = 0.583333.$$

If we now compute the product between these two numbers, we get:

$$x_m \cdot y_m = 0.428571 \cdot 0.583333 = 0.249999.$$

Thus, the computer solution has a round-off error of $e_{roundoff} = 10^{-6}$.

4.6 How do we deduce the equations used to predict the weather and the climate?

Predicting the weather and the climate involves knowing the evolution of the atmosphere and the ocean, which are two fluids: thus, we deduce the prediction equations by applying the laws of physics to these two fluids.

Starting from the laws of physics applied to a fluid, we can write a set of equations that describe the key processes that determine the main weather phenomena. This set of equations is deduced by approximating some terms of the equations that would be too complex to be solved even numerically. They are then solved numerically, and thus the solutions are affected by round-off errors. As computer power increases, some of the approximations can be gradually relaxed, and new "relevant" processes are included. In general, we say that although a set of equations provides an approximate description of reality, they are "fit for purpose" if they provide good-quality forecasts.

Forty years ago, when computer power was about a factor 10^6 less than today, the equations of motions that could be solved were extremely simple. For example, they included only a simplified description of moist processes and did not include a parameterization of the deep soil, and their representation of the interaction between radiation and the clouds was very crude. Furthermore, the mesh onto which the equations were solved had a grid point every 250–500 km, and the vertical extension of the simulated atmosphere was limited and included less than 10 vertical levels. As computer power increased, the parameterizations have been improved, and more processes have been included in the models. For example, today the state-of-the-art weather prediction models also include a dynamical representation of the ocean currents, ocean waves, sea ice, and vegetation, and the cloud schemes include different types of meteors that interact among themselves and with radiation. The radiation schemes themselves, which describe how radiation propagates within the atmosphere, have been made more accurate. As computer power continues to increase, all parameterizations will be further improved, and processes that have not yet been included in the models today but that we know can play a role in determining the weather will be introduced (e.g., the impact of the carbon cycle on the concentration of greenhouse gases in the free atmosphere, or the impact of aerosols on radiation and clouds' formation).

The equations of physics from which we start to deduce the equations are the ideal gas equation, the hydrostatic approximation, the laws of thermodynamics, the equations of kinematics, Newton's laws of motion, and the conservation of mass:

- *The ideal gas equation*—All gases follow approximately the same equation of state that links pressure, density, and temperature: by applying this equation, if we know two of these variables, we can deduce the third one.
- *The hydrostatic approximation*—The force of gravity plays an important role in determining the motions in the atmosphere, and the hydrostatic equation states that for a small mass of air, in first approximation, gravity and the pressure exercised by the air above and below the mass of air are in balance. As a consequence, we can deduce a relationship between the variation of pressure in the vertical and the air density. This relationship brings

a simplification to the equation of motions, which can be written in a format that is easier to solve, while remaining realistic enough to describe the main features of the atmospheric motions.

- *The first law of thermodynamics*—The first law of thermodynamics is a statement of the principle of conservation of energy. Consider a volume of air, which has a macroscopic kinetic energy (linked to its average motion) and has an internal energy (linked to its temperature, i.e., the microscopic motions of the molecules, and to its potential energy). The first law of thermodynamics states that, if this volume of air takes a certain amount of heat, it will either change its internal energy, or it will exercise some work against the neighboring masses of air. By applying the first law of thermodynamics, for example, we can compute the effect of the incoming solar radiation on the state of each volume of air of the atmosphere, on their temperature and macroscopic motion. We can also compute the effect that the condensation or evaporation of water has on the state of a volume of air.

- *The second law of thermodynamics*—The second law of thermodynamics is concerned with the maximum fraction of heat that can be converted into work (e.g., done by a small volume of air against its environment). This law is applied, for example, to compute how the state of a mass of air that includes water vapor evolves as it moves vertically, cools down, and releases condensed water as precipitation.

- *The equations of kinematics*—These equations are applied to describe motions, and they describe how the equations of motions change as we change the system of reference. Kinematics describes the properties of a flow, and the equations of kinematics determine, for example, how we can compute the vorticity and divergence of a flow starting from its velocity components. With appropriate choices of the system of reference used to study motions, the equations of motions can be written in simpler formats that can be more easily understood and solved.

- *Newton's laws of motion*—The first law of motion states that an object will not change its motion unless a force acts on it; the second law states that the force on an object is equal to its mass times its acceleration; and the third law states that when two objects interact, they apply forces to each other of equal magnitude and opposite direction. These three laws are applied to deduce the equation of motions of a fluid.

- *The conservation of mass*—The conservation of mass states that the mass of a volume of fluids must be conserved. It is used, for example, to derive an equation that relates how the density of air within a fixed volume changes under different types of wind field (e.g., with a shear, diverging or converging).

Other equations and relationships are used in the Earth system models to relate variables. For example, they are used to compute how a cloud is formed, how the waves of the ocean interact with the low-level atmospheric wind, how precipitation is absorbed by the soil and is stored in the soil layers, or how different types of vegetation lead to different fluxes of energy and humidity from the soil to the atmosphere. All these further equations are deduced by applying conservation laws and the laws of thermodynamics, and laws and equations of chemistry when they involve interactions with different chemical species (e.g., when we aim to model the carbon cycle).

4.7 What is an example of a set of equations that can predict the weather?

A set of equations that have been widely used to study the weather and the climate are the "primitive equations": they include all relevant processes necessary to build a model capable of predicting the weather.

The primitive equations are a set of basic equations that govern the evolution of large-scale motions, and that can be used to predict reasonably well the weather and study the climate. Although deducing the equations is beyond the scope of this text, we think it could be interesting to show them and describe their meaning.

The starting point from which the equations are deduced is Newton's second law of physics:

$$F = m a. \tag{4.1}$$

where F stands for the three-dimensional force (a vector), m for mass, and a for the three-dimensional acceleration (a vector), and the energy and mass conservation equations for a fluid (see Holton and Hakim 2012; Hoskins and James 2014).

Following Holton and Hakim (2012), the dynamic equations for a unitary fluid volume of the atmosphere on a rotating sphere (the Earth) can be written in the following form:

$$\frac{dv}{dt} = -2\,\Omega \times v - \frac{1}{\rho}\,\nabla p + g + P_v \quad \text{(momentum equation)} \tag{4.2}$$

$$c_v \frac{dT}{dt} + p\frac{d\alpha}{dt} = P_T \quad \text{(thermodynamic energy equation)} \tag{4.3}$$

$$\frac{dq}{dt} = P_q \qquad \qquad \text{(water vapor conservation)} \tag{4.4}$$

$$\frac{1}{\rho}\frac{d\rho}{dt} + \nabla \cdot v = 0 \quad \text{(continuity equation)} \tag{4.5}$$

$$\frac{dp}{dz} = -\rho g \qquad \text{(hydrostatic balance)} \tag{4.6}$$

where:

- v is the two-dimensional horizontal wind vector, $v = (u,v)$.
- Ω is the Earth angular velocity vector, directed along the Earth's rotation axis
- ρ is the atmosphere density

- p is the pressure
- $\mathbf{g}$ is the gravity vector, with magnitude and directed toward the Earth's center
- T is the temperature
- R is the gas constant for dry air ($= 287\ \mathrm{J\ kg^{-1}\ K^{-1}}$)
- c_v is the specific heat at constant-volume ($= 717\ \mathrm{J\ kg^{-1}\ K^{-1}}$)
- $\alpha = \dfrac{1}{\rho}$ is the specific volume
- q is the specific humidity
- $\mathbf{P}_v$, $\mathbf{P}_T$, and $\mathbf{P}_q$ are the tendencies of, respectively, the horizontal wind components, temperature, and specific humidity, due to physical processes (such as convection and the interaction of clouds with radiation)

These are differential equations, where "differential" means that they express a relationship between infinitesimal changes in time and in space of the variables, represented by the derivatives in space (∇) and in time $\left(\dfrac{d}{dt}\right)$. The equations are expressed in terms of six atmospheric state variables: the two horizontal wind components (the vertical velocity component is deduced using the hydrostatic equation and is not an independent variable in this set of equations), density, temperature, surface pressure, and specific humidity q. Vertical velocity is not needed, since it can be deduced following the hydrostatic approximation.

Four of the primitive equations [(4.2)–(4.5)] are prognostic and are used to compute the variation in time of the state variables, expressed as the time derivative of the variables, as a function of the forces acting on the fluid. Let's take, for example, the momentum equation (4.2):

$$\frac{dv}{dt} = -2\Omega \times v - \frac{1}{\rho}\,\nabla p + g + P_v \quad (\text{momentum equation}) \quad (4.2)$$

It states that the velocity field v changes in time if the sum of all the terms on the right-hand side is not null. The first term represents the Coriolis force, an apparent force due to the fact that the system of references used in these equations rotates with the Earth. The second term represents the pressure gradient force: it says that if a volume of air is subjected to a gradient of pressure (e.g., if the fluid on the left of the volume induces a stronger pressure than the fluid on the right of the volume), it will change its velocity. The third term represents the impact of the force of gravity, and the fourth term represents the effect of physical processes on the wind field.

The other prognostic equations are used to compute how temperature, the humidity content of a volume of air, and the air density evolve in time, while the prognostic equation (4.6) is used to relate changes in the vertical direction of pressure to differences in the air density. This "hydrostatic" equation, used in conjunction with the ideal gas equation and the continuity equation, can be used to estimate vertical velocity.

The terms (Pv, P_T, P_q) on the right-hand-side of equations (4.2), (4.3), and (4.4) represent the effect of physical processes on the state variables, or more precisely on their derivatives, that is, how they change in time. For example, they include the impact of convection on energy and moisture fluxes, the effect of clouds on the absorption, refraction, and reflection of short-wave solar radiation and long-wave radiation emitted by the Earth's surface, the effect of mountains on atmospheric flows, and the effect of turbulence on the energy and momentum transport. They are the most difficult terms to be defined and computed, and they are also one of the key sources of model approximations and forecast errors.

4.8 How do we solve the primitive equations?

The primitive equations are solved numerically, using supercomputers, on a three-dimensional grid that covers the whole globe.

The primitive equations are solved numerically on a grid, a three-dimensional mesh that covers the atmosphere (a similar approach is followed for the ocean). The resolution of this grid, the distance between each point, varies and depends on the application. If, for example, the equations are solved for numerical weather prediction valid for the next 10 days, we need to use a grid such that the numerical solution can be found in a reasonable amount of time, let's say ideally in a few hours; otherwise it would become obsolete by the time it is available. Today, the grid used to make global predictions has a resolution of about 10–25 km. If instead we are solving the equations for a research project, then the time it takes to complete the computation can be longer. In case of climate studies, the numerical integrations span a very long period, perhaps a few decades instead of a few days, or even a few weeks or months, and if we want to have a numerical solution in a reasonable amount of time (say, in about 30 days), we have to use a courser grid. Indeed, today, climate predictions are generated using grids with a resolution of about 50–200 km.

The horizontal spacing varies, for global models, from about 9 km (for the ECMWF high-resolution model version) to about 200 km (for models used for climate projections). Vertically, these models use between a few dozen to about 150 levels, which usually span the first 40–80 km of the atmosphere. In the ECMWF global model, for example, the number of vertical levels is 137, defined to span the atmosphere up to about 80 km.

The equations are integrated in time using finite-difference methods, on very powerful supercomputers. The finer the resolution, the higher the number of grid points onto which the equations must be solved, and thus the bigger the number of floating point operations that must be completed to solve the equations. Thus, computer power availability is one of the key elements that determine the model resolution.

The second element that determines the choice of the model resolution is the scales and forecast range that are the main focus of the forecasting system. If one aims to predict phenomena characterized by a very fine resolution (e.g., wind gusts linked to the passage of a frontal system across a mountain range characterized by complex orography), the model needs to be able to "resolve" the relevant scales involved in the phenomena (although it is true to say that some of the scales can be parameterized or simulated using stochastic methods). In this case, to keep the number of grid points to a reasonable amount, one can limit the region within which the model equations are solved. These types of models are called "limited area models": a typical example is the models used by the most advanced national meteorological services to generate forecasts on a grid with a spacing of about 2–3 km for a region that covers their national interest for up to 48 or 72 hours. By contrast, if one aims to predict slowly varying, large-scale patterns, a coarser resolution could be sufficient.

The third element that determines the model resolution is whether the forecasting system aims to provide not only a forecast of the most likely scenario but also an estimate of the forecast confidence, for example, expressed in terms of a range of possible forecast scenario. In this case, one needs to devote computer power to generate not only one but an ensemble of numerical integrations (usually between 10 and 50), which can be used to compute confidence intervals around the most likely scenario (e.g., defined by the ensemble mean), and probabilities that different weather scenarios could occur. For example, global ensembles have a resolution which is about a factor of 2–4 coarser than the single, global high-resolution models.

4.9 What is the difference between a prognostic and a diagnostic equation?

A prognostic equation allows us to compute how the state variables that describe a system evolve in time, while a diagnostic equation states a relationship between different variables at the same time.

Diagnostic equations are used, for example, to compute how radiation is reflected and scattered by air masses with different physical characteristics (density, water vapor content, temperature, macroscopic velocity), and thus to deduce the latter from the former, which is measured by the instruments flying onboard satellites.

Diagnostic equations are also used to derive variables of interest for weather users starting from the model state variables. For example, from the atmospheric conditions, and in particular the wind field, one can compute the wind gusts at each grid point, using equations that link the two. Similarly, depending on the atmospheric conditions, in particular the water vapor content, temperature, and vertical velocity, one can compute the type of precipitation (rain, snow, hail).

If we consider the primitive equations (4.2)–(4.6), the first four are prognostic equations while equation (4.6) is a diagnostic equation. The fact that the primitive equations are mainly prognostic mean that they predict the time evolution of the atmosphere from an initial time t_0 to future times t. The hydrostatic equation (4.6) is used to compute vertical velocity.

4.10 Can we use simple models to understand the behavior of complex systems?

Simple models can be used to understand relationships between different phenomena, but they cannot provide all the details that complex, state-of-the-art Earth system models can simulate.

A hierarchy of models of decreasing complexity, and thus of decreasing computation cost, are used to study the weather and the climate. The state-of-the-art Earth system models used to generate numerical weather predictions and to study the climate are the most complex ones: they have been designed to simulate in a realistic way all relevant processes. They are not perfect, but they are capable of capturing the key features, and to predict small-scale weather features up to few days ahead. Developing them requires a team of very capable scientists, who know the physics and chemistry behind all the key relevant processes, and are capable of writing software that simulates these processes in a realistic way. They also require a huge number of observations, to assess their quality and diagnose their performance, and to compute the initial conditions from which the prognostic equations are integrated numerically to generate forecasts. Finally, they require supercomputers, powerful enough so that their solutions can be generated in a reasonable amount of time (say, a very few hours for operational weather prediction). Thus developing, maintaining, and using a state-of-the-art model is extremely challenging, and in fact there are a limited number of institutions that develop and use them successfully, such as the top-quality national weather prediction centers and research institutes.

Their complexity makes it sometimes difficult to understand why they fail, and how different phenomena, or scales, interact and lead to changes in the weather patterns. Thus, sometimes we use simplifier models, based on a simpler set of equations, with which it is easier to understand the relationships between different processes, and that are faster to integrated numerically. For example, while a global numerical model solves the primitive equations on a 10 km resolution horizontal grid with 137 vertical levels, simplified versions with a 100–250 km grid and a few dozen vertical layers can be used to investigate issues such as the impact of large-scale surface temperature anomalies in the tropics on the general circulation of the atmosphere. Models can be further simplified to having a horizontal resolution of 250–1,000 km and less than 10 vertical levels if one is interested in assessing the evolution of even larger-scale patterns, such as, for the example, the role of the large-scale mountain chains as the Himalayas or the Andes on the atmospheric flow, or how signals and errors propagate between the two hemispheres.

When these simplified models are used to study a scientific problem, it is essential that they are not simplified too much, and that they are able to simulate realistically the key processes that can play a role in the problem under investigation; otherwise they will not be able to give relevant information on how the more complex, and realistic, models behave. Thus, deciding which simplified model versions can be used to understand the behavior of more complex models and the real Earth system, requires careful thoughts, as well as and carefully designed numerical experimentation.

At the far end of the complexity spectrum are models with a very few degrees of freedom that can be used to understand general features of the Earth system, such us the chaotic behavior of the atmosphere due to the non-linearity of its processes. For example, they can be applied to investigate predictability issues such as the role that improvements in the estimation of the initial conditions, or of model improvements, or both can have on forecast error growth.

4.11 Key points discussed in Chapter 4 "Modeling the Earth system"

These are the key points discussed in this chapter:

- Physics is at the basis of our understanding of the Earth system and of the models used to simulate its behavior.
- The state variables of a system are the variables used to describe its state and evolution: for the atmosphere, the state variables used in most of the models are temperature, density, wind, surface pressure, and humidity.
- The Earth system has an extremely large number of degrees of freedom, on the order of 10 billion (10^{10}).
- Equations describe the relationships between different variables: the simplest ones can be solved analytically (with pencil and paper) while the complex ones can only be solved numerically.
- The most commonly used set of equations that describe the Earth system are the primitive equations: they can be solved numerically on a finite mesh that covers the

whole atmosphere and the ocean, and they can be used to produce good-quality weather forecasts and climate projections.

- The key difference between prognostic and diagnostic equations is that the former describe how variables evolve in time, while the latter ones describe relationships between variables at a fixed time.

- A hierarchy of models is often used to understand the behavior of complex systems.

5

NUMERICAL
WEATHER PREDICTION

In this chapter we discuss how we solve numerically the primitive equations and generate weather predictions, and we illustrate the key steps involved in this process. More specifically, we will be addressing the following questions:

1. How do we solve numerically the primitive equations?
2. What are the key steps involved in operational weather prediction?
3. How do we determine the initial conditions?
4. What is data assimilation?
5. Do we need a supercomputer for numerical weather prediction?
6. Do we need an Earth system model to predict the weather?
7. What are the key differences between a global and a limited-area model?
8. How can we assess whether a model is realistic and accurate?
9. How much data are involved in weather prediction?

5.1 How do we solve numerically the primitive equations?

The primitive equations are solved numerically on a supercomputer, by replacing the derivatives with finite difference, from the initial time to the (final) forecast time.

We can illustrate this procedure by considering the momentum equation (4.2), simplified here to include on the right-hand side only the pressure gradient term, and assuming that we can neglect vertical velocity when we expand the left-hand-side total derivative. The wind vector is expanded into its two components $v = (u, v)$, so that we can deduce the two equations that explain how the wind changes along the x-axis (that describes motions along a parallel, from west to east) and along the y-axis (that describes motions along a meridian, from south to north). Thus, in this example we simplify the equation (4.2) into the following one:

$$\frac{dv}{dt} = -\frac{1}{\rho}\nabla p. \tag{5.1}$$

Let us now expand both the time total derivative and the pressure gradient force in their components along the two horizontal (x- and y-) directions:

$$\frac{dv}{dt} \equiv \left(\frac{du}{dt}; \frac{dv}{dt}\right) \equiv \left(\frac{\partial u}{\partial t} + u\frac{\partial u}{\partial x} + v\frac{\partial u}{\partial y}; \frac{\partial v}{\partial t} + u\frac{\partial v}{\partial x} + v\frac{\partial v}{\partial y}\right). \tag{5.2a}$$

$$-\frac{1}{\rho}\nabla p \equiv \left(-\frac{1}{\rho}\frac{\partial p}{\partial x}; -\frac{1}{\rho}\frac{\partial p}{\partial y}\right). \tag{5.2b}$$

This leads to two equations, one for each direction of motion:

$$\frac{\partial u}{\partial t} = -u\frac{\partial u}{\partial x} - v\frac{\partial u}{\partial y} - \frac{1}{\rho}\frac{\partial p}{\partial x}. \tag{5.3a}$$

$$\frac{\partial v}{\partial t} = -u\frac{\partial v}{\partial x} - v\frac{\partial v}{\partial y} - \frac{1}{\rho}\frac{\partial p}{\partial y}. \tag{5.3b}$$

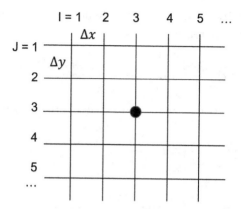

Figure 5.1. Schematic of a two-dimensional grid used to solve numerically the equations of motion (see text for more details).

These two equations are solved on a mesh that covers the whole globe (Figure 5.1): in each point of this mesh, given the value of the right-hand-side terms, we can compute how the wind components change in time.

Let us use the two indices (i = 1,2, . . .) and (j = 1,2, . . .) to represent each point position with respect to the x- and y-coordinate system, and indicate with Δx and Δy the distance between two points in the x- and y-directions, and with Δt the time difference between one time step and the next. Let us also denote with $u(x,y;t)$ the value of the u-component of the velocity at position (x,y) and at time t: thus, for example, $u(3,3;0)$ denotes the value of the u-component of the wind at the grid point (3,3) at time t = 0.

Suppose that we are the first time step, t = 0, and that we want to integrate in time the two equations by one forecast step. Also suppose that, thanks to observations covering the whole grid, we know the value of all the state variables in each point at t = 0.

We can transform the two differential equations (5.3a,b) into finite difference equations, by replacing the differentials with finite differences. Different methods can be used to transform a differential equation into a finite difference one: here, for simplicity, the time derivative will be approximated with a "forward" difference:

$$\frac{\partial u}{\partial t}(x,y;t) \approx \frac{\Delta u}{\Delta t}(x,y;t) = \frac{u(x,y;t+\Delta t) - u(x,y;t)}{\Delta t}. \quad (5.4)$$

and any space derivative with a centered difference:

$$\frac{\partial u}{\partial x}(x,y;t) \approx \frac{\Delta u}{\Delta x}(x,y;t) = \frac{u(x+\Delta x,y;t) - u(x-\Delta x,y;t)}{2\Delta x}. \quad (5.5)$$

$$\frac{\partial p}{\partial x}(x,y;t) \approx \frac{\Delta p}{\Delta x}(x,y;t) = \frac{p(x+\Delta x,y;t) - p(x-\Delta x,y;t)}{2\Delta x}. \quad (5.6)$$

If we apply these approximations, we can replace each differential with a finite difference and transform the equations (5.3a,b) into the following ones:

$$
\begin{aligned}
\frac{u(x,y;t+\Delta t) - u(x,y;t)}{\Delta t} \\
= -u(x,y;t)\frac{u(x+\Delta x,y;t) - u(x-\Delta x,y;t)}{2\Delta x} \\
- v(x,y;t)\frac{u(x,y+\Delta y;t) - u(x,y-\Delta y;t)}{2\Delta y} \\
- \frac{p(x+\Delta x,y;t) - p(x-\Delta x,y;t)}{2\Delta x}
\end{aligned}
\quad (5.7a)
$$

$$
\begin{aligned}
\frac{v(x,y;t+\Delta t) - v(x,y;t)}{\Delta t} \\
= -u(x,y;t)\frac{v(x+\Delta x,y;t) - v(x-\Delta x,y;t)}{2\Delta x} \\
- v(x,y;t)\frac{v(x,y+\Delta y;t) - v(x,y-\Delta y;t)}{2\Delta y} \\
- \frac{p(x,y+\Delta y;t) - p(x,y-\Delta y;t)}{2\Delta y}
\end{aligned}
\quad (5.7b)
$$

These two equations can be solved to compute the value of the wind at time $t + \Delta t$, as a function of the values of the state variables at time t:

$$
\begin{aligned}
u(x,y;t+\Delta t) \\
= u(x,y;t) + [-u(x,y;t)\frac{u(x+\Delta x,y;t)-u(x-\Delta x,y;t)}{2\Delta x} \\
- v(x,y;t)\frac{u(x,y+\Delta y;t)-u(x,y-\Delta y;t)}{2\Delta y} \\
- \frac{p(x+\Delta x,y;t)-p(x-\Delta x,y;t)}{2\Delta x}]\Delta t
\end{aligned}
\tag{5.6a}
$$

$$
\begin{aligned}
v(x,y;t+\Delta t) \\
= v(x,y;t) + [-u(x,y;t)\frac{v(x+\Delta x,y;t)-v(x-\Delta x,y;t)}{2\Delta x} \\
- v(x,y;t)\frac{v(x,y+\Delta y;t)-v(x,y-\Delta y;t)}{2\Delta y} \\
- \frac{p(x,y+\Delta y;t)-p(x,y-\Delta y;t)}{2\Delta x}]\Delta t
\end{aligned}
\tag{5.6b}
$$

Let us now apply these two equations to compute the wind field at point (3,3) at time $t = 10$ min, given that we know the wind, pressure, and density field at $t = 0$. For density, let us assume that it is constant over the whole grid, $\rho = 1 kgm^{-3}$, while the wind and pressure fields are shown in Figure 5.2. Let us assume that the grid spacing is $\Delta x = \Delta y = 50\,km$, and that the time step of the integration is $\Delta t = 10\,minutes = 600\,s$. Note that all variables in the equations have to be expressed in the correct units, which are meters for distances, seconds for time, m/s for velocities, kg/m³ for density, newton for force (where 1 newton is equal to 1 kg m/s²), and pascal for pressure (where

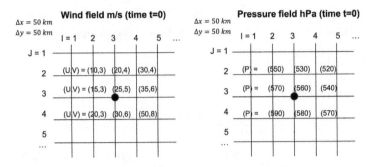

Figure 5.2. Wind (left panel) and pressure (right panel) fields at time $t = 0$, at the nine grid points centered on point (3,3) (identified by the black dot): the wind is expressed in meters per seconds, and pressure in hectopascals (1hPa is 100 pascals; 1 pascal is the pressure that a force of 1 newton exercises on a surface of 1 m²).

1 pascal is the pressure that a force of 1 newton exercises on a surface of 1 m²).

Let us now compute the value of the u-component of the wind at point (3,3) at time $t = 10$ min, first by identifying the correct points to compute the finite differences:

$$u(3,3;600s)$$
$$= u(3,3;0) + [-u(3,3;0)\frac{u(4,3;0) - u(2,3;0)}{2 * 50000}$$
$$- v(3,3;0)\frac{u(3,2;0) - u(3,4;0)}{2 * 50000} \qquad (5.7a)$$
$$- \frac{p(4,3;0) - p(2,3;0)}{2 * 50000}] 600.$$

$$v(3,3;600s)$$
$$= v(3,3;0) + [-u(3,3;0)\frac{v(4,3;0) - v(2,3;0)}{2 * 50000}$$
$$- v(3,3;0)\frac{v(3,2;0) - v(3,4;0)}{2 * 50000} \qquad (5.7b)$$
$$- \frac{p(3,2;0) - p(3,4;0)}{2 * 50000}] 600.$$

and then by replacing the variables with their numerical values:

$u(3,3;600s)$

$$= 25 + \left[-25 \frac{35-15}{2*50000} - 5 \frac{20-30}{2*50000} - \frac{540-570}{2*50000} \right] 600. \quad (5.7a)$$

$v(3,3;600s)$

$$= 5 + \left[-25 \frac{6-3}{2*50000} - 5 \frac{4-6}{2*50000} - \frac{530-580}{2*50000} \right] 600. \quad (5.7b)$$

we can calculate:

$$u(3,3;600s) = 25 + [-25*0.12 + 5*0.06 + 0.18]$$
$$= 25 - 2.52 = 22.48. \quad (5.8a)$$

$$v(3,3;600s) = 5 + [-25*0.018 + 5*0.012 + 0.3]$$
$$= 5 - 0.09 = 4.91. \quad (5.8b)$$

Thus, if you were an observer sitting at the grid point with coordinate (3,3) of the area shown in Figure 5.1, and at time $t = 0$, you had observed a wind with components ($u = 25, v = 5$) m/s, after 10 minutes, due to the action of pressure and to the advection of the masses of air, you would have observed a change of (−2.52,−0.09) m/s, which would have changed the wind into ($u = 22.48, v = 4.91$) m/s.

Although this is an extremely simple example, it should have clarified what it means to solve the primitive equations numerically. Note that, for each wind component, we made about 10 numerical calculations (where a calculation is defined as a difference, a ratio, or a multiplication). Note also that when a model is integrated over the whole globe, these calculations have to be done at each time step, and at each grid point. Let us now estimate how many calculations are involved in a 15-day forecast.

Table 5.1 reports an estimate of the number of grid points of the atmospheric three-dimensional grid and an approximate estimate of the number of calculations required to complete

Table 5.1 Estimates of the total number of grid points of a three-dimensional grid covering the globe, with a horizontal resolution of 50 km and 100 vertical levels, and estimate of the total number of floating point calculations required to complete a 15-day forecast with the primitive equations (4.2)–(4.6)

Earth radius (r; km)	6,370	
Horizontal resolution ($\Delta x, \Delta y$, km)	50	
N_x: number of points on a parallel at the equator $\left(\dfrac{2\pi r}{\Delta x}\right).$	800	
N_y: number of parallels $\left(\dfrac{2\pi r}{2\Delta y}\right).$	400	
Number of points at on a two-dimensional surface ($N_x{}^*N_y$)	320,000	
N_L: number of vertical levels	100	
NG: number of points on the three-dimensional model mesh ($N_x{}^*N_y{}^*N_L$)		32,000,000
Forecast length in minutes (15 days)	21,600	
N_T: number of time steps (with a 10-minute time step)	2,160	
N_E: number of equations	6	
N_c: number of calculations per grid pointer time step	10	
N_I: number of instructions per calculation	5	
Total number of calculations ($N_G{}^*N_T{}^*N_C{}^*N_E{}^*N_C{}^*N_I$)		20,760,000,000,000

a 15-day forecast with a model with a 50 km horizontal resolution and 100 vertical levels. The estimate is based on the following assumptions:

- Horizontal resolution of 50 km and 100 vertical levels.
- Time step of the integration of 10 minutes.
- The forecast is generated using the primitive equations (4.2–4.6), that is, six independent equations (two momentum equations, the thermodynamics energy equation, the water vapor equation, the continuity equation, and the hydrostatic balance).

- Each equation at each grid point and time step requires ten floating point calculations (sums, multiplications, divisions, exponential, . . .), and each of these calculations, once coded, involves five instructions.

From Table 5.1 we read that, with these assumptions, the number of grid points is 32 million, and the total number of floating point calculations is about 20 teraflops (1 teraflop is 10^{12} floating point calculations). For a resolution of 10 instead of 50 km, and 137 instead of 100 vertical levels, the number of grid points would be about 11 billion (instead of 32 million), and the number of floating point calculations would be about 23,000 teraflops.

5.2 What are the key steps involved in operational weather prediction?

The four key steps involved in operational weather prediction are as follows: (a) collect in a timely manner all available observation covering the area of interest; (b) generate good-quality initial conditions merging in an optimal way all the collected observations and a first-guess state; (c) integrate numerically the equations of motion; and (d) generate and disseminate weather forecast products.

Figure 5.3 is a schematic of the numerical weather prediction (NWP) process followed every day at the leading operational weather prediction center, ECMWF, to generate global forecasts:

- As many observations as possible are collected from a dedicated, global telecommunication network.
- A few times a day (for global prediction, today this happens usually every 6 hours, at times coinciding with what are called the synoptic times: 00, 06, 12, and 18 UTC, where UTC stands for Coordinated Universal Time), a data assimilation procedure is performed to estimate

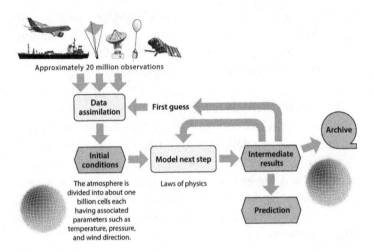

Figure 5.3. Schematic of the numerical weather prediction (NWP) process followed at operational weather prediction centers such as the ECMWF to generate a global forecast. (Source: ECMWF)

the state of the system and generate what is called the "analysis."

- At the end of the data assimilation procedure, a numerical weather forecast is generated starting from the analysis.
- Once the forecast has been completed, forecast products are generated and disseminated.
- Note that with the term "first guess," we mean the first few hours (say, up to 12) of a forecast, which are used by the data assimilation procedure as an estimate, a first guess, of the state of the atmosphere up to the next analysis time.

Necessary conditions to complete these steps successfully and produce good-quality forecasts involve having a critical mass of scientists who know the physics of the Earth system and how to solve numerically the fluid equations in a very efficient way; a good-quality weather forecast model; a good-quality data assimilation system; and a supercomputer.

5.3 How do we determine the initial conditions?

Initial conditions are computed by merging all available observations and a first guess of the state of the atmosphere, usually defined by the most recent forecast, by applying a procedure called data assimilation.

Suppose that we have a good model based on the primitive equations (4.2)–(4.6) and a computer fast enough to complete a 15-day model integration in a reasonable amount of time (say, 1–2 hours). The first step that we need to complete to generate a numerical weather forecast is to estimate the initial conditions of the system: we need to know the state of all variables (wind, temperature, pressure, humidity, density) and the forcing fields (e.g., the incoming solar radiation) at all grid points to be able to compute the forces acting on the air masses and how much energy is exchanged between the different Earth system components.

Schematically, we can write the primitive equations in the following way:

$$\frac{dx}{dt} = F(x,t) \qquad\qquad (5.9)$$

where here with the symbol x we denote all the state variables, and with the term F we denote the forces acting on the system. The equation (5.9) states that the infinitesimal variations in time of the state variables, $\frac{dx}{dt}$, depend on the forces, which depend themselves on the state of the system and on time.

Equation (5.9) can be integrated analytically for extremely simple systems, or numerically. Formally, we can write its solution in the following way:

$$x(t) = x(t=0) + \int_0^t F[x,t]\,dt. \qquad\qquad (5.10)$$

In the example of section (5.1) we solved this equation numerically for one time step and computed:

$$x(t) = x(t=0) + F\left[x(t=0), t=0\right] \cdot \Delta t. \tag{5.11}$$

The time integration was possible because we knew the initial state (the wind and the pressure at $t = 0$), and thus we could compute the two terms on the right-hand side of the equation, $x(t=0)$ and $F\left[x(t=0), t=0\right]$.

A similar procedure is followed when we have to integrate a global weather prediction model. All available observations are collected and merged with a "first-guess" estimate of the state of the atmosphere, using a very complex procedure named "data assimilation." We need to merge the observations with a "first guess" because, as we have discussed in Chapter 3, observations do not cover the whole globe and/or are of different qualities. The best way to fill in the gaps of the observations is to merge the available observations with a first guess, which is usually defined by the most recent short-term forecast.

Suppose that it is 15.00 UTC (Universal Coordinated Time, i.e., the Greenwich Mean Time) on May 14, we have just finished receiving all global observations covering the time period from 9.00 to 15.00 UTC, we have access to the 15-hour forecast issued at 00.00 UTC, and we want to issue a forecast that starts at 12.00 UTC and is valid for the next 15 days.

At each grid point and for each time step from 00.00 to 15.00, we have access to a short-range forecast, and thus we know a "first-guess" value of the state of the atmosphere. Since the first guess is a forecast, it is affected by errors, which can be corrected by comparing it with the observations that cover the assimilation time window, which we have collected. This comparison can be done at all the grid points where we have both the forecast and the observations: given that also the observations are affected by errors, data assimilation does not replace the model state with the observed state, but combines the two values in an optimum way. More precisely, the initial state is defined by a weighted mean of the two values, where the weights given to the model forecast and the observations depend on their relative accuracy. Data assimilation determines the "optimum weights" to give to the first guess and the observations dynamically: in other words, it does not give the same weights every day, but it computes weights that vary according to how many and which observations are available, and to the atmospheric flow. In situations that are known (by looking at the statistics of past forecast errors) to be more difficult (easier) to handle by the forecast model, more (less) weight is given to the forecast and less (more) to the observations.

5.4 What is data assimilation?

With the term "data assimilation" we mean the process followed to estimate the initial state from which Earth system models are integrated in time to generate a weather forecast.

At the time of writing (2022), every day about 600 million observations are collected worldwide and exchanged in real time (about 95% are taken by instruments onboard satellites). Of these, about 10%–15% are used to estimate the initial conditions required to start a numerical integration of the equation of motions. These observations are selected so that they provide a good and uniform coverage of the whole globe. Some of the 600 million collected observations are discarded because they do not satisfy quality control checks, while others are discarded because they are redundant. A thinning process is also applied to make sure that the spatial resolution of the observations is comparable to the model grid everywhere on the globe: this thinning step is performed also to make sure that everywhere and at every time the same spatial and temporal scales are resolved by the model and by the observations.

Observation quality is very important, since it influences the accuracy of the best estimate of the true state of the atmosphere that we generate using data assimilation procedures. The observation quality depends on the instrument, and for satellites also on the instrument position with respect to the area under observation. Satellite observation quality is influenced by the state of the atmosphere, with, in general, lower (higher) observation errors in cases of clear-sky (cloudy) conditions. For example, since water vapor concentration is higher close to the surface, satellite observations are usually less accurate in cloudy conditions closer to the Earth's surface. Furthermore, when many observations are used together to estimate the state of the atmosphere, we must also take into account their representativeness error, which depends on where they are located with respect to all the others.

Data assimilation takes into account the observation errors. To give an example of the amplitude of observation errors and how they depend on the observation instrument, Table 5.2 lists the estimated root-mean-square errors for observations of the two horizontal wind components (u and v) and temperature, at different vertical levels (from A. Geer, ECMWF, 2017, personal communication). By assigning a different error to each observation, the data assimilation procedure gives more (less) weight to the more (less) accurate observations. The most sophisticated assimilation procedures can also take into account the fact that observations taken closely to each other have correlated observation errors.

There are further complications that we should consider in data assimilation. One is linked to the fact that the model grid and the location where observations are taken, in general, do not coincide. Another one is due to the fact that the times when observations are taken might not coincide with an available model state, and a further one is due to the fact that the variables used by the model to define the state of the atmosphere (temperature, wind, humidity, surface pressure, density, and cloud concentration) do not coincide with the observed

Table 5.2 Estimated root-mean-square errors for observations

Vertical level (hPa)	U and V wind component observations (m/s)			Height observations (m)			Temperature observations (K)		
	TEMP / PILOT	SATOB	SYNOP	TEMP / PILOT	SYNOP (manual land)	SYNOP (auto land)	TEMP	AIREP	SYNOP (land)
1000	1.80	2.00	3.00	4.30	5.60	4.20	1.40	1.40	2.00
850	1.80	2.00	3.00	4.40	7.20	5.40	1.25	1.18	1.50
700	1.90	2.00	3.00	5.20	8.60	6.45	1.10	1.00	1.30
500	2.10	3.50	3.40	8.40	12.10	9.07	0.95	0.98	1.20
250	2.50	5.00	3.20	11.80	25.40	19.05	1.15	0.95	1.80
100	2.20	5.00	2.20	18.10	39.40	29.55	1.30	1.30	2.00
50	2.00	5.00	2.00	22.50	59.30	44.47	1.40	1.50	2.40

Columns 2–4: root-mean-square errors of the U and V wind components for three types of observations at seven different heights (from the surface at 1,000 hPa to 50 hPa), used in the ECMWF data assimilation procedure. Columns 5–7: as columns 2–4 but for height observations. Columns 8–10: as columns 2–4 but for temperature. SYNOP, TEMP, and PILOT are different types of observation reports from surface station; SATOB are satellite observations; and AIREP are airplane observations. (Data from A. Geer, 2017, ECMWF, personal communication)

variables. To address these complications, data assimilation uses "observation operators," pieces of software that map a model state into an observation state, and vice versa. These operators have been developed for each observation type: note that, since they simulate the true physical phenomena only in an approximate way, they can also introduce uncertainties in the way the initial conditions are estimated.

Another complication is linked to the fact that the first-guess errors depend on the state of the Earth system itself. To address this flow dependency, data assimilation gives a flow-dependent weight to the first guess, estimated, for example, by using an ensemble of first guesses (i.e., an ensemble of short-range forecasts).

Having a very good estimate of the initial conditions is essential to generate top-quality forecasts, and the most advanced operational meteorological centers have developed their own assimilation method to generate them. These methods can be grouped into three main categories: variational methods; ensemble methods; or hybrid methods, which combine an ensemble component, used to provide flow-dependent statistics, and a variational component. (For the reader who wants to know more about these methodologies, see the Further Reading chapter at the end of this book.)

5.5 Do we need a supercomputer for numerical weather prediction?

Yes, we do. Without a computer we would not be able to integrate numerically the equations that we need to solve to generate a weather forecast: with a slow computer we would be able to solve only a simplified version of the equations, while with a supercomputer we can integrate a state-of-the-art Earth system model and generate top-quality forecasts.

A supercomputer is just a very fast computer, capable of completing trillions of floating point operations in a second. Table 5.3 lists the performance of a few, top-quality supercomputers used in meteorology and climate science, compiled using entries from the "Top-500 supercomputers" list of June 2022. As reference points, we have also listed the performance of the 1st, the 2nd, and the 500th machines. For each host institution, Table 5.3 lists the ranking of the machine, its name and manufacturer, the installation site, the number of processors (cores), the maximal LINPACK performance achievement (R_{max}), and the theoretical peak performance R_{peak}. When an installation site has more than one machine, we have listed all its machines and ranked it accordingly to the R_{max} of its top machine.

The maximal LINPACK benchmark is a measure of the computer's performance when it is used to solve a set of linear equations $A \cdot x = b$, where A is an $N \times N$ dimensional matrix, and x and b are N-dimensional vectors. Since in the LINPACK benchmark the same test is performed on all computers, it allows us to compare in a clean way their performance. But we should be aware that the top performance in solving more complex problems (e.g., in generating a weather forecasts) could be less than R_{max}. R_{peak} is the top performance indicated by the computer manufacturer, a measure that depends on the applications used by them to measure it.

From Table 5.3 we see that in June 2022, the Korea Meteorological Administration (KMA) supercomputers top the meteorological installations, with two identical machines ranked 31st and 32nd, each with R_{max}(KMA) = 18.00, with about 300,000 processors, and a combined peak performance R_{peak} (KMA) of 51 petaflops (1 petaflop is 10^{15} floating point operations per second). KMA is followed by Meteo France, the UK Meteorological Office, the German Climate Center, the Japanese Meteorological Administration (JMA), ECMWF, the German Meteorological Centre (DWD), the Indian Institute of Tropical Meteorology (IITM), and the United States National Oceanic and Atmospheric Administration (NOAA).

Note that only the top machine, Frontier, reaches an exaflops (10^{18} flops) R_{max}, and only the first three machines have an R_{max}

Table 5.3 List of some of the top supercomputers used in operational weather prediction and climate centers

Tp500 rank	Machine	Institution—Country	Cores	Rmax (PFlops/s)	Rpeak (PFlops/s)
1	Frontier (HPE Cray EX235a)	DOE/SC/Oak Ridge National Laboratory—United States	8,730,112	1,102.00	1,685.65
2	Fugaku	RIKEN Centre for Computational Science—Japan	7,630,848	442.01	537.21
31 + 32	Guru and Maru (2* ThinkSystem SD650 V2)	Korea Meteorological Administration—South Korea	612,864 (2*306,432)	36.00 (2*18.00)	51.00 (2*25.50)
71	Belenos (Bull Sequana XH2000)	Meteo France—France	294,912	7.68	10.47
75 + 207 + 208	3*Cray XC40	UK Meteorological Office—UK	421,632 (241,920 + 2*89,856)	12.64 (7.04 + 2*2.80)	14.17 (8.13 + 2*3.02)
76 + 179	Levante and Mistral (Bull Sequana XC2000 and Bullx DLC 720)	Deutsches Klimarechenzentrum—Germany	317,998 (308,096 + 99,072)	10.01 (7.00 + 3.01)	14.24 (10.28 + 3.96)
94 + 95	2*Cray XC50	Japan Meteorological Agency	271,584 (2*135,792)	11.46 (2*5.73)	18.26 (2*9.13)
109	Cheyenne—SGI ICE XA	National Center for Atmospheric Research—United States	144,900	4.79	5.33

Column 1	Column 2	Column 3	Column 4	Column 6 (R_{max})	Column 7 (R_{peak})
128 + 129	2*Cray XC40	European Center for Medium-Range Weather Forecasts—UK	252,936 (2*126,468)	7.84 (2*3.94)	3.78 (2*1.89)
130 + 155	SX-Aurora TSUBASA A412-8	Deutscher Wetterdients—Germany	33,024 (18,688 + 14,336)	7.12 (3.87 + 3.25)	9.89 (5.61 + 4.28)
132	Pratyush—Cray XC40	Indian Institute of Tropical Meteorology—India	119,232	3.76	4.01
176	Hera—Cray CS500	NOAA Environmental Security Computer—United States	63,840	3.08	4.88
500	Software Company (M) A1	Lenovo Hosting Services—United States	57,600	1.65	2.12

Column 1 reports the "Top-500" ranking of the machine in June 2022; column 2 reports the machine name and manufacturer; column 3 reports the installation site; column 4 reports the number of processors (cores); column 6 reports R_{max}, the maximal LINPACK performance achievement; and column 7 reports R_{peak}, the theoretical peak performance. In the "Top-500" list, the computers are ranked by their R_{max} value. When an installation site had more than one machine, we have listed all its machines and ranked it accordingly to the R_{max} of its top machine. (Data from the "Top-500" list of June 2022, available at https://www.top500.org/lists/top500/2022/06/)

above 100 petaflops. Note also that the KMA machine has a performance that is about 3% of the top machine, Frontier. The meteorological installed machines have an R_{max} between 3 and 36 petaflops and a ranking between 30 and 200, which can be seen as an indication that weather and climate applications are among the top users and consumers of computer power.

Today, supercomputers are based on an architecture that involves thousands of processors. When we integrate numerically an Earth system model on one of these machines, different regions of the world are assigned to different processors, and time-step computations are performed simultaneously on all the allocated processors. At the end of a time step, data are exchanged among the processors, so that at the next time step all processors have all the data that they need to make the next computations. This means that a substantial part of the available computational time has to be devoted to communication.

To reduce the total number of central processing units (CPUs) required to complete a numerical weather forecast, the model software is written in a such a way that the computation is distributed in the most efficient way among all the processors. Note that the optimal computation configuration is not to devote as many processors as possible to the numerical integration, since this would increase dramatically the time required to exchange data among processors, but to identify an optimal number of processors, so that a large amount of calculations can be done in parallel on different processors without needing to spend too much time for communication.

5.6 Do we need an Earth system model to predict the weather?

There is not a clear answer to this question, since it depends on the phenomena we want to predict, on their spatial and temporal scale, and on the time range up to which we want to predict them. For example, we definitely need a model that includes a three-dimensional dynamical simulation of the ocean and of sea ice if we want to issue predictions valid for longer than 1 week, while for forecast ranges shorter than 3 days we could not include them.

Suppose that "Operational Prediction Center A" is running a coupled (dynamical ocean and atmosphere) 10 km model to predict high-intensity precipitation events and wind storms up to 72 hours ahead. In other words, "Operational Prediction Center A" is focusing on including all the relevant processes that can affect the evolution of weather phenomena over a 72-hour period, and on being able to simulate in a realistic way the local, small and fast scales. Since the ocean and sea ice evolve on time scales slower than 72 hours, they could replace a fully three-dimensional dynamical ocean with a mixed-layer scheme that is capable of simulating the daily cycle of the ocean surface temperature, and replace a dynamical sea-ice model with a static scheme that does not vary in sea-ice cover during the 72-hour integration. They could then devote the "saved" computer time to increase the resolution of the Earth system components that are kept in the model: the atmosphere and the land surface.

By how much could they increase the resolution of the model by not including a dynamical ocean and sea-ice model? Suppose that the model has a horizontal resolution of 10 km, and that the coupled three-dimensional dynamical ocean and dynamical sea-ice models add 50% to the computer time required by the atmosphere-and-land model components. If they reduced the grid spacing of the atmosphere-and-land model components by 20%, from 10 to 8 km, they would increase the computer cost of integrating them by 50%. In fact, using an 80% finer model resolution would increase the number of points by a factor $1.25 = 1/0.80$ in each horizontal direction, thus lengthening the computer cost by a factor $1.56 = (1.25)^2$. This is equivalent to adding 56% to the computing cost, a value that is close to the 50% saving linked to the exclusion of the dynamical ocean and sea-ice model. Suppose that experiments have indicated that a resolution increase from 10 to 8 km leads to detectable improvements in high-impact weather forecasts in the short range. Then, they could decide to make the change: un-couple the model and increase the resolution from 10 to 8 km.

Consider now "Operational Prediction Center B," which is running an uncoupled 25 km model to generate large-scale pattern (e.g., temperature) predictions in the 2-to-4-week fore-cast range. In other words, "Operational Prediction Center B" is focusing on the monthly forecast range. Suppose that they know that a 2-to-4-week forecast is sensitive to changes in the ocean and sea-ice state, and that not including them makes the simulations of the interactions between the ocean and the atmosphere less realistic, and the forecast errors larger. "Operational Prediction Center B" could decide to save com-puter time by using a coarser resolution and devote the saved computer time to include a three-dimensional dynamical ocean and sea-ice model. They could decrease the resolution from 25 to 30 km: this would lead to a reduction of the number of points by a factor $0.83 = 25 / 30$ in each horizontal direction, which would decrease the computer cost to $0.69 = (0.83)^2$. If they then coupled the 30 km model to a three-dimensional dy-namical ocean and sea-ice model, the computer time would increase by 50%, that is, from 0.69 to $1.035 = 0.69 \cdot 1.5$. Thus, in this case the saving linked to the reduction of the resolu-tion would pay for the inclusion of the dynamical three-dimensional ocean and sea-ice model.

A third, very interesting case is linked to the prediction of hurricanes. Their development, intensification, and path are determined by the interaction of a vortex that develops in the at-mosphere with the underlying ocean. As a hurricane develops and propagates over the sea, it extracts energy to sustain it-self, and as a consequence, the temperature of the sea water below its track cools. This means that if a hurricane moves very slowly and stays for many hours over the same area of the sea, it would cool it down, and in the long term it would not be able to extract the same heat as before to sustain itself, and thus lose power. To accurately predict a hurricane's development, intensification, and propagation, the model needs to be able to simulate this dynamical interaction between the ocean and the atmosphere: simple mixed-layer ocean models could already provide a detectable improvement compared to keeping the sea surface temperature constant, although it would be best to couple the atmosphere to a three-dimensional dynamical ocean. This is especially true if one wants to extend the hurricane prediction to lead times longer than a few days, say to 5–15 days, and it is essential if one wants to predict the probability that intense storms could develop in an ocean basin in the forthcoming weeks, in sub-seasonal forecasts.

These examples illustrate how operational meteorological centers define the configuration of their operational models: depending on their main objectives, and in particular on the scales and time range of their numerical model integrations, they decide how best to distribute the computational resources they have available, whether to use them to include more Earth system components, to include more sophisticated schemes that simulate some physical phenomena, or to increase the model resolution.

5.7 What are the key differences between a global and a limited-area model?

A limited-area model covers only a limited region, instead of the whole globe. It is used to generate very high-resolution, short-range forecasts covering a limited area of interest. Compared to a global model, to integrate numerically a limited-area model, one needs not only initial but also boundary conditions.

Computer power availability limits how many floating point operations can be performed in a reasonable amount of time, say in about 2 hours, and thus it limits the number of grid points and time steps that can be set in a numerical weather integration. The main reason why operational prediction centers use limited-area models is that, if they are mainly tasked to produce short-range (say 2-to-3-day) forecasts over a specific region, given a certain amount of available computer resources, they could limit the area of integration and use a very high resolution inside this area.

A limited-area model simulates the same physical processes as a global model, but usually at a higher resolution than a global model. Compared to integrating numerically a global model, there is an extra complication when integrating in time a limited-area model: since it covers a small region, and since weather propagates, one needs to update the state of the atmosphere at the boundary of the model using forecast fields produced by a global model. Boundary forecast updates are required to make sure that during the time integrations, the correct weather systems propagate into the area of interest.

Suppose that we want to generate high-resolution forecasts over central Europe, say over an area that covers the Euro-Atlantic sector, for a 72-hour period. Suppose that we have access to global forecasts at a 10 km resolution valid for the whole forecast period, with a 3-hour time frequency, and that we want to use a 2 km resolution over the limited area, while keeping the same vertical resolution as the global model within the troposphere, but limit its vertical extension to about a quarter the one of the global model.

Let us estimate the relative computer power cost of this limited area forecast compared to a global forecast. The grid of this limited-area model would have in the horizontal 25 grid points every 1 global grid point, and say half of the number of vertical levels. Since the resolution is 5 times finer, the integration time step has to be smaller, say by a factor of 4. Thus, within the area of interest the limited area would have 12.5 (25/2) more grid points, while in time each global time step would be covered by 4 time steps of the limited-area model. This makes the limited-area model integration about 50 times (12.5 times 4) more expensive. If the limited-area model covers an area that is about 10 times smaller than the surface of the Earth, then the total computer cost of a 3-day limited area forecast with a 2 km resolution would be comparable to the cost of a global 15-day forecast with a 10 km resolution.

This is how we should proceed to integrate the limited-area model:

- *Initial conditions*—They are needed to start the numerical integration. We can simply extrapolate them from the global fields on the finer grid of the limited-area model. If we have a data assimilation scheme capable of assimilating extra data with a finer resolution than the ones used in the global model, we could use the global fields as a first guess, and then run the data assimilation procedure with also the extra observations to generate high-resolution initial conditions.
- *Boundary conditions*—They are needed to update the atmospheric flow at the boundary. We can define them using global forecasts valid every 3 hours, from the initial time to the end of the integration time, over a frame around the limited area of integration with a width of few grid points, and extrapolate them to the limited area finer resolution.
- *Numerical integration*—This is performed as with a global model, on a limited-area finer grid and with a shorter time step.

As the forecast time progresses, the limited-area model generates local, higher-resolution and faster scales that the global model is not able to simulate. Thus, it should, for example, be better capable to simulate the effect of the mountains on the atmospheric flow and to localize more precisely and more realistically moist events (clouds, thunderstorms).

As weather phenomena propagate into the area from the incoming boundary, the limited-area model adjusts them and introduces the smaller scales that the global fields were not capable of simulating. The finer scales included in the numerical integration have an impact on the large-scale fields, and this should lead, on average, to improved forecasts also for the large scales. In other words, using a limited-area model would not only generate forecasts that include finer details, smaller spatial scale, and faster waves, but should also lead to improved forecasts of the larger scales.

We are using the conditional tense because the limited-area model could perform worse than the global model, if, for example, the physical parameterisations that it uses are of an inferior quality to the ones used in the global model. In other words, higher-resolution does not necessarily lead to better forecasts.

Care should be taken in selecting where to put the boundary with respect to the area of interest. Ideally, the boundary should be put far enough so that weather events entering the area have enough time to develop and adjust before they reach the area of interest for the whole time integration.

5.8 How can we assess whether a model is realistic and accurate?

Models are assessed by comparing their performance against reality, both on single test cases, selected to assess the performance of some specific model components, and statistically on a large number of cases (ideally covering a whole year).

Numerical weather prediction models include different components (land, ocean, atmosphere, and sea ice) and many parameterization schemes (e.g., schemes that simulate clouds, vegetation, turbulence, radiation, air-sea interaction), which all need to be tested. These tests should include an assessment of the quality of the numerical schemes used to integrate numerically the model equations and of data assimilation.

A model assessment includes usually a first phase, during which each parameterization scheme is assessed individually, both on single individual cases and on a large number of them, and a second phase when the performance of the whole model is assessed on many cases, covering at least a warm and a cold season, and ideally one full year.

Suppose that we are in the process of upgrading the operational model (the o-suite, where "o" stands for operation) with a new version (the e-suite, where "e" stands for experimental) that includes both upgrades of existing model components (e.g., convection in the atmosphere and sea ice) and new components (e.g., a dynamic vegetation scheme). To decide whether the e-suite is ready to replace the o-suite in operational forecast production, first each individual component is assessed individually: the validation procedure includes some specific case studies, selected because they allow for testing the most relevant aspects of the new component (e.g., whether some driving parameters have been set to the correct number), and a statistical evaluation on a large number of cases (say, a few dozen). This second statistical assessment is required to be sure that the upgraded scheme behaves well, on average, in any weather situation, and not just on a few selected weather conditions.

Once each single component has passed its individual validation phase, the upgrades are all merged into the e-suite model version, which is tested against the o-suite. This second testing phase usually includes many more cases than the first phase, at least a warm and a cold season, and ideally 1 year of cases. This phase is essential to avoid the potential overall negative impact on forecast quality that changes in some model components could have when used in conjunction with all the other upgrades. The performance of the e-suite and the o-suite is assessed both against observations and analyses (generated by merging observations and first-guess fields generated using the e-suite model in data assimilation). To provide a thorough assessment and the entire three-dimensional flow, many different fields are considered, and different verification metrics are applied.

At the end of this comparison, scorecards are produced to summarize the relative performance of the e-suite versus the o-suite in different seasons, geographical areas, atmospheric levels, variables, and forecast steps. If, overall, the scorecards indicate a positive impact, the new model version is implemented in operation, and the e-suite becomes the new o-suite. If the scorecards indicate a negative impact, more testing and diagnostic work is performed to understand the reasons of the negative performance and to prepare a new e-suite version.

5.9 How much data are involved in weather prediction?

A huge amount of data! The most comprehensive weather data archives include a few hundred petabytes (i.e., 2^{50} bytes, or 1,024 terabytes): these include past observations, analyses, and forecasts.

There are at least four reasons why national meteorological centers have archives that contain huge amount of data, observations, analyses, and forecasts that span many years:

- *Benchmarking*—New model versions are often tested also on old, interesting weather events (e.g., major failures, or extreme weather events that caused major damages), to assess how the state-of-the-art prediction system would have performed in the past. This is possible only if past data are kept.
- *Evaluating progress*—Operational weather prediction centers are often asked to evaluate progress in forecast accuracy. Sometimes new evaluation methods are developed, or variables that had not been evaluated in the past have now become of interest. A backward evaluation is possible only if we can extract from the archive the old forecasts, and compare them with the quality of the most recent ones.
- *Generating calibrated products*—In the past two decades, new methods have been developed to generate better, more accurate weather forecasts by calibrating them. Calibration schemes usually need a few decades (20–35 years) of past forecasts. In the future, it is expected that artificial intelligence (AI) methods could help design new calibrated forecast products: the longer the training datasets that we will be able to provide to these methods, the better these AI-calibrated products could be.
- *Assessing climate trends*—To assess climate trends, many decades of observations and analyses are required. As climate change has been accelerating and more and more countries have been impacted by climate change, the interest has been moving from variables such as

temperature and precipitation to other variables such as wind, sea-level height, soil moisture, solar radiance, river discharge, vegetation, ice, and snow cover. Observations, analyses, and reanalyses (analyses of past decades generated using today's state-of-the-art models and data assimilations) of all these variables are extremely valuable resources for climate studies, and they should be kept in the archives to allow more comprehensive trend analyses.

We can get an estimate of how much data are produced daily by considering ECMWF: they collect all available observations daily, generate analyses four times a day (every 6 hours) and ensemble forecasts up to 15 days, generate twice-a-week ensemble monthly forecasts valid for up to 46 days, and generate once-a-month ensemble seasonal forecasts valid for up to 7 months. On average, in 2020, they were adding daily to their archive about 200 terabytes. In 2020, they also reported that their data archived contained about 270 petabytes (1 petabyte is equal to 2^{50} bytes, or to 1,024 terabytes; 1 terabyte is equal to 2^{40} bytes); thus, at the time of writing (August 2022), the ECMWF archive probably contained about 400 petabytes.

As a reference, consider that in August 2022 the top-of-the-line laptops included a hard drive that contained about 1 terabyte. Thus, in August 2022 the ECMWF archives contained the equivalent of about 400,000 top-of-the-line laptops!

5.10 Key points discussed in Chapter 5 "Numerical weather prediction"

These are the key points discussed in this chapter:

- The primitive equations are solved numerically, using supercomputers, by replacing differentials with fine differences.

- Numerical weather prediction includes four main steps: collecting observations; estimating the state of the atmosphere and computing the initial conditions; integrating numerically the equation of motion; and generating weather forecast products.
- Initial conditions are determined using data assimilation that merges in an optimal way all available observations and a first-guess field, usually provided by the most recent, short-range forecast.
- Given the scale and complexity of weather prediction, we need top-performing supercomputers to be able to generate good-quality weather forecasts in a reasonable amount of time.
- Given a certain amount of available computing resources, there is always a trade-off between dedicating computing resources to increase model complexity or model resolution; the choice of the operational configuration of a numerical prediction suite depends on its objectives, the scales, and the time range it aims to predict.
- Limited-area models are used to zoom in on an area of interest and generate short-range (say 48- to 72-hour) forecasts that include finer details than a global forecast.
- Model realism and accuracy are assessed by comparing model forecasts with reality, both on a few specific cases of interest and on a large number of cases covering at least two seasons, ideally 1 year.
- Meteorological centers host huge data archives that contain observations, analyses, and forecasts that span several decades, for benchmarking new model versions, evaluating model progress, generating products, and assessing climate trends.

6

CHAOS AND
WEATHER PREDICTION

In this chapter we discuss the chaotic nature of the atmosphere, how small, local errors can grow and affect a weather forecast globally, and how we can estimate forecast uncertainty using ensemble methods. More specifically, we will be addressing the following questions:

1. What is a chaotic system?
2. What is Lorenz's three-dimensional model?
3. What is the "butterfly effect"?
4. What are the sources of forecast error?
5. How can we reduce initial condition uncertainties?
6. How can we reduce model uncertainties?
7. How do we measure forecast errors?
8. What is an ensemble?
9. Are ensemble forecasts more valuable than single ones?

6.1 What is a chaotic system?

A dynamical system shows a chaotic behavior if most orbits exhibit sensitive dependence, which means that most other orbits that pass close to it at some point do not remain close to it as time advances.

This definition is general and applies to any dynamical system. Consider an N-dimensional system, that is, a system with N state variables. The state of the system at any time t can be identified by a point in the N-dimensional space defined by its state variables, its "phase space." If we now consider the state of the system at consecutive times, and join them with a continuous line, we can draw the orbit of the system. In the case of a 3-dimensional system, one orbit is the line in the 3-dimensional phase space of the system that joins all the points that the system goes through as time progresses, starting from its initial point.

Now, consider a large set of initial conditions (large enough that it spans the whole phase space of the system) and compute the system's orbits starting from all these points for a very long forecast time. Once we have performed these integrations, we have populated the phase space of the system with the system's orbits. The definition of a chaotic system reported above means that, if we consider two orbits that pass infinitesimally close at a certain time t_0, they will not remain close as time progresses, and at one point in time t_1 they will diverge.

Now suppose that we want to make a forecast of the evolution of a chaotic system from time t_0. To be able to do so, we need to estimate the system initial state at t_0. Suppose that, due to observation errors, our initial conditions estimate is uncertain. This means that the real state of the system is different from our estimated initial conditions: in other words, the two points that represent our estimated initial conditions and the real initial conditions do not coincide. Since the system has a chaotic behavior, the two orbits that are very close at t_0 will at one point in time diverge, and our forecast will be affected by an error. This error can grow and at a certain time in the future make our forecast no better than a climatological estimate or a random forecast.

The atmosphere exhibits this behavior. Sometimes people talk of the atmosphere as a classical example of "deterministic chaos," in the sense that the trajectory divergence is not linked to model errors, or to the fact that we are missing some terms in the model equations, or that terms in the equations might be uncertain, stochastic in nature. Even if we know perfectly well the equations of motion, and we can integrate them, two "deterministic" solutions that pass infinitesimally close to each other at one point in time would eventually diverge.

6.2 What is Lorenz's three-dimensional model?

Edward Lorenz's three-dimensional model is a simplified, nonlinear model of convection in the atmosphere, which exhibits a chaotic behavior and has been used to investigate predictability.

Edward (Ed) Lorenz is considered one of the fathers of chaos theory. By comparing the behavior of the real atmosphere and his three-dimensional simplified version, he was able to clarify concepts such as atmospheric predictability, and to understand the role that small errors in one part of the globe could play on forecasts valid for the whole globe. His experiments and his scientific paper that discussed how "one flap of a seagull's wings could change the course of weather" led to the definition of the concept of the "butterfly effect."

Ed Lorenz's model was inspired by the works of others, as can be traced back to his articles and looking at the references therein. In particular, it is based on the earlier works of John William Strutt (Lord Rayleigh) and Barry Saltzman.

John William Strutt, Lord Rayleigh (1842–1919), in 1916 wrote a paper in the *Philosophical Magazine and Journal of Science*, where he investigated convection currents in a horizontal layer of fluid (Figure 6.1). Starting from Newton's laws applied to this fluid, he deduced a simplified set of equations. Although he could not solve them analytically, he applied them to explain the results of experiments performed at that time. This was just one of his works, not the one that made him famous. In fact, in 1904 he received a Nobel Prize in Physics for his investigations of the densities of the most important gases and his discovery of argon.

Barry Saltzman (1931–1969) started from the equations deduced by Lord Rayleigh and also applied them to investigate convection currents in a horizontal layer of fluid. He

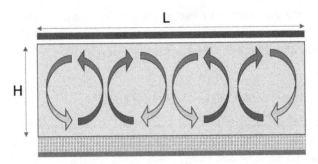

Figure 6.1. Schematic of the system studied by Lord Rayleigh in 1916. A liquid in an infinitely long channel is heated from below and cooled from above. The two-dimensional system has a characteristic horizontal length L and a characteristic vertical height H. Heating from below induces a warming of the lower layers, which triggers vertical motions, with warmer rising fluid replaced by colder descending fluid.

looked for some special solutions that included only certain scales and ended up with a set of 52 ordinary nonlinear equations. Among them he deduced a highly truncated system with only seven components, which he said were sufficient to describe the key characteristics of convection. Of the seven equations, he then identified three that, alone, would capture the key characteristics of the flow: two components represented the flow motion (expressed in terms of its streamline) and the thermal fields of a critical mode of instability, and a third component represented the departure of the vertical temperature stratification from the initial linear variation. He said that these three variables, and relative equations, would be able to capture the key features of the flow convection. Also for Saltzman this was just one of his works: in fact, he is known mainly for his contribution to our understanding of the past climate and its oscillations with a 100,000-year cycle. For this and other contributions, in 1998 he received the Carl Gustaf Rossby Research Medal of the American Meteorological Society.

Edward Lorenz started from the equations deduced by Saltzman and noted that of the simplified seven equations, all but three would tend toward zero for long forecast times. These three equations coincide with the three components that Saltzman already had identified as capable to describe the flow key characteristics. He rewrote these three equations by redefining the three state variables and arrived at the following formulation:

$$\begin{cases} \dfrac{dX}{dt} = -\sigma X + \sigma Y \\[2mm] \dfrac{dY}{dt} = -XZ + rX - Y \\[2mm] \dfrac{dZ}{dt} = XY - bZ \end{cases} \qquad (6.1)$$

$$\sigma = \frac{v}{k}; b = \frac{4}{\left(1+a^2\right)}; r = \frac{R_a}{R_c}; R_a = \frac{g\varepsilon H^3 \Delta T_0}{kv}; R_c = \frac{\pi^4}{a^2}\left(1+a^2\right)^3$$

$$(6.2)$$

where σ is the Prandtl number, v is the coefficient of viscosity, k is the coefficient of thermal conductivity, $a = H/L$, R_a is the Rayleigh number, and R_c is a critical Rayleigh value.

Except for multiplicative constants, the three state variables (X,Y,Z) are the same as Saltzman's three key variables: X is proportional to the intensity of the convection motion; Y is proportional to the temperature difference between the ascending and descending currents; and Z is proportional to the distortion of the vertical temperature profile from linearity.

Despite their simplicity, Lorenz's three-dimensional system contains a good deal of realism and has a clear connection with more complex systems and reality. The use of the Lorenz three-dimensional system to investigate predictability in the real atmosphere that has at least 10^7–10^9 degrees of freedom is a very nice example of how simple systems can be used to understand the behavior of reality.

Convection is a key motion of the atmosphere that can trigger major changes in the atmospheric flow, and it is also a key motion in much simpler systems, such as a flow in a channel. If we can understand how convection forms and triggers instabilities in a simple system, like the flow in a channel, then we can apply this knowledge to the atmosphere. This is precisely what the works of Lord Rayleigh and Saltzman did: they deduced a set of simpler equations that could be treated and studied (remember that at that time computers did not exist) and investigated their behavior in some limiting, interesting, and realistic conditions. Lorenz used these equations to perform an instability analysis of the flow and understand how perturbations would grow in such a system. He then extrapolated his results to the more complex, real atmosphere.

Figure 6.2 shows two orbits, two numerical integrations for 20 seconds (2,000 time steps, with an integration time interval of 0.01 seconds) of Lorenz's equations starting from two

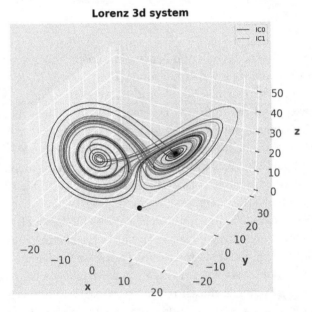

Figure 6.2. Lorenz three-dimensional system. Two trajectories, one black and one gray, starting at two very close points [with coordinates (0,1,0) and (0.05,0.95,−0.05)], identified by the black circle, are shown in the three-dimensional phase space of the system for 20 seconds (2,000 time steps, with a time interval of 0.01 seconds). The two trajectories overlay for about 6 seconds (600 time steps) and then start diverging. Close to the end of the time integration, say after 15 seconds, the black trajectory remains in the right wing of the attractor, while the gray trajectory keeps oscillating between the left and the right wings.

very close points with coordinates (0,1,0) and (0.05,0.95,–0.05). These two points are superimposed and identified in the figure by the single black circle.

These two integrations, named orbits in the language of chaos theory, have been generated by setting the parameters in equation (6.1) to $\sigma = 10$, $r = 28$, $b = 8/3$, $dt = 0.01$, and by integrating the equations numerically in time for 2,000 time steps. The two trajectories overlay for about 6 seconds (600 time steps) and then start diverging. Close to the end of the time integration, say after about 15 seconds (1,500 time steps), the black trajectory remains in the right wing of the attractor, while the gray trajectory keeps oscillating between the left and the right wings. Figure 6.3 shows the same two trajectories of

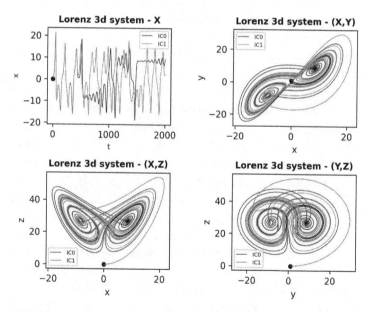

Figure 6.3. Lorenz three-dimensional system. The same two trajectories of Figure 6.2 are shown here in four different two-dimensional cross sections. The top-left panel shows an (X; time) cross section: for each time t, the figure shows the X value of the two trajectories, $X_1(t)$ and $X_2(t)$, for the first 2,000 time steps. The other three panels show the two trajectories in the three two-dimensional cross sections defined by the (X,Y), (X,Z), and (Y,Z) coordinates.

Figure 6.2 in four different, two-dimensional cross sections. The top-left panel show an (X; time) cross section: for each time t, the figure shows the X value of the two trajectories, $X_1(t)$ and $X_2(t)$, for the first 2,000 time steps. The other three panels show how the two trajectories look in the three two-dimensional cross sections (X,Y), (X,Z), and (Y,Z); that is, if we projected the image shown in Figure 6.3 on a plane identified by two of the three coordinates.

One of the two initial points can be considered as our best estimate of the "true" state of the system, and the other one as the true state, and thus their difference represents the initial condition error. This example shows that, even with very small initial errors, at one point in the forecast time the two trajectories start diverging, and the forecast error starts growing. In the language of chaos theory, the two orbits that at one point in time (the initial time) pass very close to each other do not remain so in the future.

These two figures show that the orbits tend to fill in two areas of the phase space of the system that resembles the wings of a butterfly. The area spanned by all the system's orbits defines the system attractor. The two wings of the Lorenz attractor can be considered as two different weather regions—for example, one hot and dry without convection, and the other one colder and wet with convection. Suppose that we are aiming to predict whether the weather could turn hot and dry, and cause drought conditions. While it is still possible to predict with a good degree of accuracy the future forecast state of the system for a short time (6 of the 20 seconds we integrated the equations for), it is difficult to predict whether the system will turn hot and dry in the long forecast range. Similar orbits' sensitivity to the initial state is shown in weather prediction.

Lorenz was on the faculty of the Massachusetts Institute of Technology (MIT) from 1955 to 1988. Through his profound contributions to science as well as his quiet demeanor, gentle humility, and love of nature, he set a compelling example of what it means to be a gentleman and a scholar. In discovering "deterministic chaos" he established a principle that "profoundly influenced a wide range of basic sciences and brought about one of the most dramatic changes in mankind's view of nature since Sir Isaac Newton," said a committee that awarded him the 1991 Kyoto Prize for basic sciences. He is best known for the notion of the "butterfly effect," the idea that a small disturbance like the flapping of a butterfly's wings can induce enormous consequences, although in the first paper that he published that discussed this phenomenon he talked about a seagull, and not a butterfly . . . but since Lorenz's attractor does remind us of a butterfly, in later works he talked about the "butterfly effect"!

6.3 What is the "butterfly effect"?

The notion of the "butterfly effect" is based on the idea that a small disturbance like the flapping of a butterfly's wings somewhere in the world could induce consequences to the atmospheric flow in a fine period of time.

The "butterfly effect" means that a little difference in the initial state of a system, as could be due to the difference between state (X_A, Y_A, Z_A) and the state represented by those values perturbed by a very small value, $(X_B, Y_B, Z_B) = (X_A + \varepsilon_X, Y_A + \varepsilon_Y, Z_A + \varepsilon_Z)$, can grow and lead to the two trajectories moving far apart after a certain amount of time.

Sensitivity to initial conditions is popularly known as the "butterfly effect" because of the title of a paper given by Edward Lorenz in 1972 to the American Association for the Advancement of Science in Washington, DC, entitled "Predictability: Does the Flap of a Butterfly's Wings in Brazil Set Off a Tornado in Texas?" The flapping wing represents a small change in the initial condition of the system, which causes a chain of events that prevents the predictability of large-scale phenomena. Had the butterfly not flapped its wings, the trajectory of the overall system could have been vastly different.

As suggested in Lorenz's book entitled *The Essence of Chaos*, "sensitive dependence can serve as an acceptable definition of chaos." In the same book, Lorenz defined the butterfly effect as follows: "The phenomenon that a small alteration in the state of a dynamical system will cause subsequent states to differ greatly from the states that would have followed without the alteration." The above definition is consistent with the sensitive dependence of solutions on initial conditions of a chaotic system.

A consequence of the sensitivity to initial conditions is that if we start with an incomplete set of information about the state of any system, as is usually the case in weather prediction since observations are affected by errors and do not cover all the scales that affect the weather, then beyond a certain time the system will no longer be predictable.

6.4 What are the sources of forecast error?

The two key sources of forecast error are uncertainties in the initial conditions and uncertainties in the models.

Observations are affected by errors, and they provide a limited spatial and temporal coverage. This is the first source of forecast error.

The other source is linked to the models, which provide only an approximate description of reality. Even if we had perfect model equations, the fact that they can only be solved numerically introduces errors. Furthermore, since computer resources are limited, the models only simulate some of the real physical processes, and the ones that they include are simulated in an approximate way.

If we now consider that the initial conditions used to produce weather forecasts are computed by combining observations and models in data assimilation systems, it is clear that initial conditions also are uncertain.

The combination of model and initial condition uncertainties is the source of forecast errors. The chaotic nature of the atmosphere makes the error growth flow dependent: in certain situations, errors grow slowly, and thus forecasts remain accurate for a long period. In others, errors grow more quickly, and the forecast error reaches large values in a short period .

6.5 How can we reduce initial condition uncertainties?

Initial conditions are estimated using data assimilation that merges all available observations with a first guess (i.e., short-range forecasts); thus, initial condition errors can be reduced by improving the observation network, data assimilation, and the model used to generate the first guess.

The global observation network can be improved by installing and launching more observation platforms, thus widening the observation coverage, and by adopting more accurate observation instruments. Models can be ameliorated by adding more relevant processes, improving the simulation of the processes already included, and increasing the model resolution. Data assimilation can be improved by upgrading the assimilation method, for example by adopting a stronger, more realistic coupling so that observations taken in one component of the Earth system (e.g., the ocean) can have an impact also in other components (e.g., the atmosphere), or by making the assimilation procedure more efficient so that all the collected observations are used (today, daily we use only about 10%–15% of the collected data).

A possible improvement that has not yet been implemented is the use of a strongly coupled data assimilation scheme that assimilates observations in the atmosphere, the land, the ocean, and the sea ice. Prototype coupled schemes have been developed and used to generate coupled reanalyses of the past decades, but they have not yet been used in operations weather prediction, partly because of technical challenges and partly because of their very high computing cost.

A second possible improvement would be to exploit in operation the idea of using more observations in target regions,

regions where initial errors grow faster. This could be achieved by reducing the thinning of the observations in these areas, and possibly even by collecting more satellite observations in these areas. A similar technique has been tested with conventional data in the past to collect extra targeted dropsonde observations to improve the prediction of high-intensity weather systems, and it is used routinely during the hurricane season to improve the initialization of hurricanes. To our knowledge, this technique has not yet been tested in operations with satellite data.

6.6 How can we reduce model uncertainties?

Model uncertainties are due to the fact that models simulate in an approximate way reality, either because they do not include all relevant physical processes, or because the ones that they include simulate reality in an approximate way. A further source of uncertainty is linked to the fact that models use a finite resolution. By acting on these three aspects we can reduce model uncertainties.

Since the beginning of operational numerical weather prediction in the 1980s, thanks to the work of many scientists and the continuous increase in computer power, we have been able to solve more complex model equations on finer spatial grids, to include more processes, and to improve the realisms and accuracy of the processes that were already included.

Figure 6.4 links the availability of increasing computer power with model resolution at the European Centre for Medium-Range Weather Forecasts (ECMWF). Computer power is measured in terms of sustained teraflops and is indicated on the left axes on a logarithmic scale. Resolution is indicated on the graph with a label and is indicated on the right-hand-side vertical axes in terms of the number of model grid points corresponding to each resolution. Each rectangle represents a computer machine used at ECMWF from 1980 to 2022: its width indicates the years it was used in operation, and its height indicates its sustained performance (in teraflops, as indicated on a left-hand-side logarithmic scale).

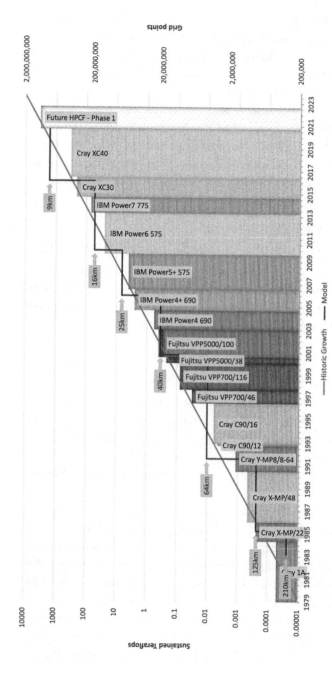

Figure 6.4. Time evolution of ECMWF computer power (left axes) and model resolution (right axes) from 1980 to 2022. Computer power is measured in terms of sustained teraflops and is plotted on the left axes on a logarithmic scale. Resolution is indicated on the graph with a label. The right-hand-side vertical axes report, as a reference, the number of model grid points on one of the model vertical layers corresponding to each horizontal resolution. (Source: ECMWF)

Note that the model's horizontal resolution was changed roughly every 5–6 years, as indicated by the labels. The right-hand-side vertical axis provides an indication of the number of grid points on a model vertical layer that corresponded to each model horizontal resolution.

The straight line in Figure 6.4 shows the linear fit of the logarithm of the sustained computational power with respect to time. The line has a slope of about a factor of 2 every 18.5 months: in other words, during the period from 1980 to 2022, on average, sustained performance (SP) increased by a factor of 2 every 18.5 months:

$$\frac{SP(Y_1)}{SP(Y_0)} = 2^{\left[\frac{12}{18.5}\cdot(Y_1-Y_0)\right]} \tag{6.1}$$

where Y_1 and Y_0 are two years, and the factor $12/18.5$ has been introduced to fit the data. For example, if we set $Y_1 = 2019$ and $Y_0 = 1991$, we find that:

$$\frac{SP(2019)}{SP(1991)} = 2^{\left[\frac{12}{18.5}\cdot 28\right]} \sim 290,000 \tag{6.2}$$

Note that the SP-growth factor that we have found is consistent with what is called "Moore's Law," which is linked to Gordon Moore, the cofounder of Intel, who observed in 1965 that the number of transistors on a microchip doubled about every 2 years, thus leading to an exponential growth of available computer power.

Table 6.1 highlights how, between 1991 and 2019, computer power has allowed ECMWF to increase the model resolution and the model complexity, to couple a land-atmosphere model to a three-dimensional dynamical ocean, and to increase the frequency of operational weather forecast production. The table shows an estimate of the extra cost required to complete a forecast in 2019 compared to 1991, during which horizontal resolution increased by a factor 14, vertical resolution by a factor of 15.2, the time step was reduced by a factor of 5,

Table 6.1 Key characteristics of the operational model used at ECMWF, cost drivers and cost factors linked to the model upgrades, and key characteristics of the operational supercomputers, in 1991 and in 2019

	1991	2019	Computational cost increase linked to the upgrade	Ratio between the 2019 and the 1985 factors
Model type	Atmosphere, land	Atmosphere, land, ocean, sea ice		
Horizontal resolution (grid spacing in km)	125	9	14 × 14	196
Vertical resolution (number of levels)	19	137	137/9	15.2
Time step of numerical integration (seconds)	3600	720	5	5
Model complexity (number and type of physical parameterizations)	1	10	10	10
Coupling to a dynamical ocean and sea ice	No	Yes	+ 30%	1.3
Forecast production frequency (per day)	2	4	2	2
Total extra cost				~ 388,000
Supercomputer	Cray YMP-8	Cray XC40		
Sustained performance	1 Megaflops	320 Teraflops	320,000	320,000

model complexity led to a cost increase of a factor of 10, and the coupling to the dynamical ocean increased the cost by 30%. Overall, in 2019 to complete a forecast, we needed about 388,000 more computer power, and computer power did increase by a similar factor of 320,000.

The continuous increase in computer power not only allowed for an increase in resolution but also for refining the physical parameterization schemes included in the models; this improved the water cycle and the effect of clouds on radiation. In the last 20 years, for example, major changes have been introduced in the land surface schemes, in the simulation of energy and water fluxes from the deep soil to the atmosphere, the ocean waves, in the cloud microphysics scheme, and in radiation.

If we look into the future, apart from continuing to improve the existing schemes, there are five areas of development that are expected to lead to model improvements:

- In the land surface scheme: improvements are expected in the simulation of the effect of cities on the low-level atmosphere and from the introduction of dynamical vegetation.
- In the freshwater cycle: improvements are expected from closing it, so that water vapor generated at the atmosphere-ocean interface is then transported in the clouds to land areas, condenses and precipitates, percolates into the soil, flows toward rivers, and finally reaches the ocean again.
- In the dynamical treatment of aerosols in the atmosphere: improvements in the simulation of the interaction of aerosols with clouds are expected to lead to reduced radiation biases.
- In the dynamical ocean model component: a reduction of the drifts of the ocean model state is expected to bring forecast error reductions in the long range (seasonal to decadal).

- In the dynamical interaction between sea ice and ocean waves: improvements are expected from including this dynamical interaction, especially in transition seasons, when sea ice is forming or melting, and can affect ocean-atmosphere fluxes.

6.7 How do we measure forecast errors?

Forecast errors are measured using a range of metrics that compare forecasts against observations and analyses.

Operational weather prediction centers routinely compare the quality of their forecasts using a variety of metrics and considering many atmospheric fields. Metrics and fields are selected to provide a complete overview of model performance. Evaluations are performed both considering individual cases, for example, when high-impact weather events of interest occurred, and statistics accumulated over months and seasons. Daily, routine evaluations are performed to monitor weather forecasts as soon as they can be verified, so that if large forecast errors are detected, the model developers are immediately informed, so that they can immediately start to identify the reasons for failure and possible changes that would have given a better forecast.

With the general term "metric," we denote a method used to measure forecast error. There is not a unique metric, or a "best" metric. Each metric is sensitive to different variables and weather conditions: by using a combination of them, we can have a more complete assessment of forecast error. A metric is applied usually to the difference between a weather forecast and a verification field: the difference can be computed at a specific location (e.g., where a meteorological observing station is located) or over a large area (e.g., a nation, or a larger area characterized by similar average weather conditions).

Different metrics capture different aspects of forecast error. For example, suppose that we are interested in making sure that a weather event (e.g., rain or no rain) is predicted in a timely fashion, while we are less concerned about the intensity of the weather event. In this case we could consider, as a verification metric, one that is sensitive to "rain" or "no rain," even if the metric is not sensitive to the precipitation amount. The opposite is the case if we are interested in verifying the quality of our forecasts in predicting extreme weather conditions, for example, a high amount of rain or very strong wind conditions. In this case, we have to adopt a metric that is sensitive to the precipitation or wind intensity.

To provide a balanced and comprehensive evaluation of forecast quality, it is very important to consider also single weather events, especially if they are linked to high-impact weather, since these are the events that cause most of the damages. But since these weather events are rare, looking only at forecast errors in these cases would provide a limited view of model accuracy. This is why assessments based on single cases must be supported by average statistical evaluations that cover many events.

A key aspect in the computation of forecast error is whether to compute it at a specific location, for example, where meteorological stations are located, or whether to consider a large area. Again, both evaluations should be performed. Care should be taken when verifying a weather forecast produced with a model on the model grid against observations taken at specific station locations. Since the model operates on a finite grid, it cannot simulate phenomena with a spatial scale that is finer than the model grid. Thus, the model cannot be expected to provide local details, weather variations that happen on spatial and temporal scales finer than the ones it can simulate. One can always generate a weather forecast valid for a point where a meteorological station is located by extrapolating the forecast conditions from the nearest grid points, and thus compute a forecast error at the station location. But care should be taken in interpreting the results, since the model forecast might not be able to correctly simulate the local details, the daily cycle of weather variables (e.g., of temperature, wind, and cloud cover), or weather conditions that are influenced by the local terrain configuration, which might not be "seen" by the model.

A comprehensive forecast evaluation procedure should include both single cases and statistical analyses based on many cases, a selection of weather variables that characterize key weather events (temperature, wind gusts, rain, cloud cover, pressure, and so on), and a range of metrics, computed considering different areas of the globe. Since a complete evaluation procedure generates a huge number of scores, we usually summarize them in a "score card" that lists all the key numbers. Sometimes we can even combine all these scores into a unique overall number, defined by a weighted average of all the scores, where the weight of each score depends on the relevance given to this particular score.

Two of the most commonly used metrics to evaluate the quality of single forecasts are the root-mean-square error and the anomaly correlation coefficient. Denote with $f\left(x_j, t\right)$ a forecast valid at grid point $x_j \in \Sigma_G$, and $v\left(x_j, t\right)$ the corresponding verification field (defined either by observations, or grid-point analyses). Denote by $\bar{f}\left(x_j\right) = f\left(x_j, t\right)$ the climatological forecast field (e.g., the average of the forecast fields $f\left(x_j, t\right)$ computed considering many different years), and by $\bar{v}\left(x_j\right) = v\left(x_j, t\right)$, the climatological value of the verification field.

The root-mean-square forecast error of the forecast within the area Σ_G of interest that includes N_G points can be computed in the following way:

$$rmse\left(f, v; t\right) = \sqrt[2]{\frac{1}{N_G} \sum_{N_G}\left[f\left(x_j, t\right) - v\left(x_j, t\right)\right]^2} \qquad (6.3)$$

The anomaly correlation coefficient measures the correlation between the anomaly of the forecast and the verification computed with respect to reference fields, such as the model climatology and the verification climatology:

$$acc(f,v;t)$$
$$= \frac{\sum_{N_G}\left[f(x_j,t)-\bar{f}(x_j,t)\right]\left[v(x_j,t)-\bar{v}(x_j,t)\right]}{\sqrt[2]{\sum_{N_G}\left[f(x_j,t)-\bar{f}(x_j,t)\right]^2\sum_{N_G}\left[v(x_j,t)-\bar{v}(x_j,t)\right]^2}} \qquad (6.4)$$

If the area is large and includes points that have very different latitudes, a weight that depends on the latitude of the grid point x_j should be included in the computation, to consider the fact that the area that each point represents differs as we move from the equator to the poles.

For any forecast f, a verification field v, and a reference forecast f_{ref} (e.g., climatology, $f_{ref} = cli$), once we have computed the score of the forecast f, $SC(f,v)$, of the reference forecast, $SC(f_{ref},v)$, and of a perfect forecast, $SC(f_{perfect},v)$, we can define the forecast skill score:

$$SK(f,v;f_{ref}) = \frac{SC(f,v)-SC(f_{ref},v)}{SC(f_{perfect},v)-SC(f_{ref},v)} \qquad (6.5)$$

The skill score is a number between 0 and 1: 0 for a forecast that performs as the reference forecast and 1 for a perfect forecast.

6.8 What is an ensemble?

An ensemble is a practical tool designed to assess the predictability of the daily atmospheric flow. It is based on a number of numerical integrations of the model equations, with each integration designed so that the whole ensemble simulates all potential sources of forecast error.

Ensembles can be used to predict the time evolution of the probability density function (PDF) of forecast states. Predicting the time evolution of the PDF of forecast states translates, in practical terms, to being able to predict the probability of intense rainfall or cold temperatures over the Euro-Atlantic region, or to estimate the probability that a hurricane will pass within a 100 km radius of a specific location. One of the by-products of ensemble prediction is the possibility to estimate the forecast skill of the best estimate of the state of the system, or, in other words, to forecast the forecast skill.

A complete description of the weather prediction problem can be stated in terms of the time evolution of an appropriate probability density function (PDF) in the atmosphere's phase space. The predicted PDF of possible future atmospheric states is a function greater than zero in the phase space regions where the atmospheric state can be, with maximum values identifying the most probable future states. Theoretically, the problem of the prediction of the PDF can be formulated exactly through the continuity equation for the probability distribution function of forecast states. Practically, however, the continuity equation can be solved only for extremely simple systems, characterized by a very limited number of degrees of freedom. This is still true today, even with close-to-exascale computer power availability, and thus, to date, ensembles based on a finite number N of numerical integrations are the only feasible way to predict the PDF beyond the range of linear error growth.

Since generating an ensemble of N forecasts costs N times the cost of generating a single forecast, and requires the development and implementation of "nontrivial" methods to simulate initial and model uncertainties, up to the 1990s all operational weather prediction centers were issuing only single forecasts.

It is the work of Edward Lorenz, and of others inspired by his research, that made it clearer how error growth is flow dependent, and the potential benefit that having a flow-dependent estimate of the forecast error could deliver. Following his pioneering work of the 1960s, research groups started investigating in the 1970s how to estimate the forecast skill and be able to identify weather conditions that are more or less predictable than the average. This led, in the 1980s, to the development of prototype ensemble systems that were implemented in operations in the 1990s.

Figure 6.5 illustrates the key difference between a single forecast and an ensemble approach to weather prediction, projected onto two dimensions. The figures show the time evolution of a single forecast (black, dashed line labelled "Single FC_0"), of an ensemble of forecasts (gray solid lines labelled "ENS FC_j"), and of reality (gray dashed line labelled "reality"), at four times: the initial time t_0 and three forecast steps, t_1, up to which the initial PDF evolves linearly, and t_2 and t_3, during which the evolution becomes nonlinear and the ellipsoid changes shape and becomes stretched along a few, key directions with the fastest growth rate. Small circles denote the position of each forecast and reality at the four time steps.

With a single-forecast approach (Figure 6.5, top-left panel), only one forecast is issued, starting from our best estimate of the real initial conditions. The gray ellipsoid shown at initial time t_0 represents our best estimate of the initial condition uncertainties, with the single forecast starting from the most likely state, which coincides with the center of the ellipsoid. The black and gray dashed lines show the time evolution at the forecast times t_1, t_2, and t_3 of the single forecast and reality. As time progresses, they diverge, and the forecast error grows. Statistical methods based on past forecast errors can be used to estimate the forecast error, but this estimate cannot take into account the flow-dependent nature of error growth: this has been represented in the figure by the fact that the ellipsoids get stretched but do not change shape; they only grow in size and rotate. The gray rectangles plotted below the X- and the Y-axes show the forecast range estimated at time t_3 using the single forecast. Note that it is not guaranteed that the forecast range

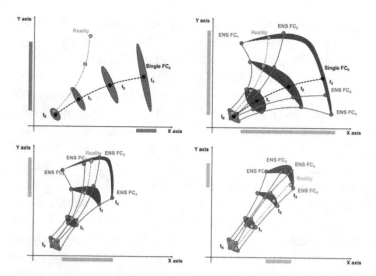

Figure 6.5. Schematic of a single-forecast (top-left) and an ensemble (top-right) approach to numerical weather prediction. The two lower panels show two other examples of ensemble forecasts. The four panels show the time evolution of a single forecast (black dashed line), of an ensemble of forecasts (gray solid lines), and of reality (light gray dashed line), as seen on a two-dimensional phase-space, at four forecast times: the initial time t_0 and three forecast steps, t_1, up to which the initial PDF evolves linearly, and t_2 and t_3. In a single-forecast approach (top-left panel), only one forecast is issued, starting from our best estimate of the real initial conditions. At initial time t_0, a gray ellipsoid represents our best estimate of the initial condition uncertainties, and at forecast times t_1, t_2, and t_3, statistical methods can be used to estimate the range of uncertainties, represented by the gray ellipsoids, at the different forecast times, but these estimates are not flow dependent. With an ensemble of forecasts (top-right panel and lower panels), it is possible to estimate how the initial-time uncertainty evolves, taking flow-dependent error growth rates into account (represented by the fact that the initial ellipsoid changes shape and is stretched along a few, key directions of error growth). The three ensembles show how the range of forecasts covered by the ensemble (the bar shown along the x- and y-axes) can be used to estimate the range of the forecast state at time t_3.

includes the projection of reality on the two axes: in fact, the X-coordinate of "reality" at final time is well below the minimum value spanned by the gray bar below the X-axis, and the Y-coordinate is above the range spanned by the gray bar below the Y-axis.

With a "reliable" ensemble of forecasts (Figure 6.5, top-right panel), we can use the N ensemble members (only four have been represented in this schematic) to estimate how the initial-time uncertainty evolves. Since each forecast evolves nonlinearly accordingly to the equation of motions, it can be used to estimate how the initial ellipsoid grows and changes shape. Thus, in this case, the gray area changes shape as the ensemble members diverge nonlinearly. With an ensemble of forecast, we can estimate how key points that describe the PDF evolve and use them to construct how the initial ellipsoid evolves. By using all N forecasts, we can estimate whether the points diverge linearly or nonlinearly, as a function of the atmospheric state. The gray rectangles plotted below the X- and the Y-axis coordinates show the forecast range estimated at time t_3 using the ensemble: note that in this case the forecast range includes the projection of reality onto the two axes: in fact, the X and the Y coordinates of reality fall within the range spanned by the gray rectangles. Let me stress the fact that it is important to have a reliable ensemble to make sure that, on average, the range spanned by the ensemble includes reality, and matches the flow-dependent predictability.

Since in the ensemble the PDF estimate depends on the atmospheric flow, in an accurate and reliable ensemble the spread of the ensemble forecasts, their divergence, can be used to estimate the possible forecast range. This can be illustrated by comparing the shapes of the gray areas, and the size of their projections onto the X and Y coordinates, in the three ensembles: while the range is wide in the first case (Figure 6.5, top-right panel), it is narrower in the second case (Figure 6.5, bottom-left panel), and even narrower in the last case (Figure 6.5, bottom-right panel). This is the power of ensembles.

It is worth clarifying what we mean with the word "reliable": we mean that, when the ensemble predicts that the possible range of forecast is very narrow (i.e., that the divergence between the individual forecast members is small), the forecast error averaged over many cases is small. By contrast, when the divergence among the ensemble members is large, the forecast error averaged over many cases is large. One way to verify the ensemble reliability is to contrast, on a two-dimensional figure, the ensemble forecast probability and the observed frequency. In this diagram, constructed considering many cases, when the ensemble gives an 80% probability that an event would occur, the event occurs 80% of the times. The diagram should show that all points defined by a forecast probability of an event and its observed frequency lie on the diagonal.

In a reliable ensemble, we can use the forecast divergence as a predictor of the forecast error. To explain why this should be the case, consider what we should expect under a "perfect model" assumption, whereby forecast errors are due to initial uncertainties only. In this case, if our ensemble of initial conditions has been defined to give a reliable estimation of the initial uncertainty, we would not be able to distinguish between an ensemble member and reality, since reality could coincide with one of them. This means that, on average, the distance of a randomly chosen ensemble member from any other should be equal to the distance of reality from any ensemble member. This implies that, if we grouped together the cases for which the average distance of a randomly chosen ensemble member and reality is relatively small (compared to all the cases), the average distance of a randomly chosen member from any other member would also be relatively small. Vice versa, if we grouped together the cases for which the average distance of a randomly chosen ensemble member and reality is relatively large, the average distance of a randomly chosen member from any other member would also be relatively large.

In other words, in a "perfect model" situation, reliable ensembles should show, on average over many cases, a correspondence between the divergence among the ensemble members and the forecast error of one of them. If the ensemble is not perfect, this relationship should still hold in a reliable ensemble, since reliability means that we have simulated all the sources of forecast error linked to initial conditions and model uncertainties.

6.9 Are ensemble forecasts more valuable than single ones?

There are two key reasons why ensemble-based, probabilistic forecasts are more valuable than single forecasts. First, an ensemble makes it possible to predict the whole distribution of possible events, and thus to assess the probability that any event could occur. Second, ensembles provide forecasters with more consistent (i.e., less changeable) successive forecasts.

The first reason why ensemble-based, probabilistic forecasts are more valuable than single forecasts is because ensembles make it possible not only to predict the most likely scenario (e.g., defined by the ensemble-mean forecast or any chosen member) but also to estimate the probability that any alternative event could occur. In other words, compared to single forecasts that give one possible future scenario only, ensembles provide users with more complete and valuable information.

One way to measure the difference in value between single and ensemble forecasts is to use the potential economic value metric (PEV). PEV is based on a simple cost-loss model: a user can decide to pay an amount (cost) C to protect against a loss L, linked to a specific weather event. The PEV can then be assessed by considering users with different C/L ratios and by constructing a curve that shows the savings that users could make if they used the forecasts. Clearly, PEV is a function of forecast quality: a poor (unreliable and/or nonaccurate) ensemble will not be able to outperform a good, single forecast.

The second reason why ensemble-based, probabilistic forecasts are more valuable than single forecasts is that they provide forecasters with more consistent successive forecasts. This can be assessed if one compares consecutive ensemble forecasts, issued 24 hours apart and valid for the same verification time. Results published in the scientific literature indicate that forecasts based on consecutive ensembles jump less than forecasts based on single integrations; that is, they give more consistent forecasts than the corresponding single forecasts.

6.10 Key points discussed in Chapter 6 "Chaos and weather prediction"

These are the key points discussed in this chapter:

- In a chaotic system, two trajectories that are very close at a certain time will eventually diverge; the divergence is flow dependent.
- The atmosphere is a chaotic system, and small initial condition errors can affect the forecast.
- Edward Lorenz's three-dimensional nonlinear system is a simplified model of convection; it has been widely used to understand forecast error growth and predictability.
- The butterfly effect indicates a strong sensitivity of a forecast to initial conditions: discussed by Edward Lorenz in one of his papers, it illustrates the fact that even a small perturbation caused by the flap of a butterfly's wings can grow and affect the forecast state everywhere in the globe.
- There are two main sources of forecast errors: initial condition and model uncertainties.
- Initial condition uncertainties can be reduced by improving the observation network, models, and data assimilation.
- Model uncertainties can be reduced by improving the model parameterization schemes, by adding missing processes, and by increasing the model resolution; thanks to the continuous growth in computer power, the past 40 years have seen major advances in all three areas.
- Forecast error is measured using a range of metrics applied to the difference between forecasts and verification, defined by observations and/or analyses.
- An ensemble is a practical tool designed to assess the predictability of the daily atmosphere, based on an ensemble of N numerical integrations.

- Ensemble forecasts are more valuable than single ones because they estimate the whole probability distribution of forecast states and thus can be used to estimate the probability of possible future weather scenarios; they are more valuable also because consecutive ensemble forecasts are more consistent than consecutive single forecasts.

7

DEALING
WITH UNCERTAINTIES AND
PROBABILISTIC FORECASTING

In this chapter we review how we simulate uncertainties in ensemble forecasting, what probabilistic forecasts are, and how ensembles are used to generate them. More specifically, we will be addressing the following questions:

1. How do we build an "accurate and reliable" ensemble?
2. What is a probabilistic forecast?
3. How can we communicate forecast uncertainty?
4. How can we make decisions using probabilistic forecasts?
5. What is a scenario forecast?
6. What is a cluster analysis?
7. How do we measure the accuracy and reliability of a probabilistic forecast?
8. What are reforecasts and reanalyses?
9. Why are reforecasts and reanalyses useful?

7.1 How do we build an "accurate and reliable" ensemble?

The strategy that should guide the development of an accurate and reliable ensemble is to simulate all potential sources of forecast errors and to use a realistic and accurate model to generate each ensemble forecast.

Since 1992, ensembles have been part of the operational suites at the European Centre for Medium-Range

Weather Forecasts (ECMWF) and at the National Centers for Environmental Prediction (NCEP, United States), and since 1995 at the Canadian Meteorological Center (CMC, Canada). Following their examples, in the following years many other operational weather prediction centers have developed and implemented operation ensembles.

All operational ensembles have been designed to simulate the sources of forecast errors due to initial and model uncertainties. The relative importance given to the two sources of forecast errors depends on the characteristics (e.g., spatial and temporal scales) of the phenomena that the ensembles have been designed to predict. For global ensembles aiming to predict large-scale atmospheric patterns in the short and medium forecast range (say, for forecasts valid for 3 to 5 days), research studies performed with state-of-the-art numerical models have indicated that forecast errors in the first few days are mainly due to initial uncertainties, and that model errors start having an impact comparable to the effect of initial errors after forecast day 3. For the prediction of small-scale, low-pressure systems and associated precipitation fields, model errors can be as important as initial uncertainties already at forecast day 2, and in some specific cases even earlier.

In general, there is not a clear separation between the impact of the two sources, or a clear forecast time for which forecasts of certain scales are more influenced by initial or model uncertainties. Furthermore, we should take into account that since models are used in data assimilation procedures to compute the initial conditions, model uncertainties affect initial uncertainties.

In the first version of the ECMWF ensemble, initial uncertainties were simulated using singular vectors (SVs), perturbations with the fastest growth over a finite time interval. Compared to random initial perturbations, SVs are characterized by a faster growth rate, which mimics the growth rate of forecast errors. SVs remained the only type of initial perturbations in the ECMWF ensemble until 2008, when perturbations computed from an ensemble of data assimilations (EDA) started being combined with SVs. This change was inspired by the strategy used in the Canadian ensemble (see later discussion). Today, SVs remain an essential component of the ECMWF ensemble, and they are used also in other global ensembles.

NCEP also designed their first version to simulate initial uncertainties only, but they used a different technique based on bred vectors (BVs). The BV cycle aims to emulate the data-assimilation cycle, and it is based on the notion that analyses generated by data assimilation will accumulate growing errors by the virtue of perturbation dynamics. This is because neutral or decaying errors detected by an assimilation scheme in the early part of the assimilation window will be reduced, and what remains of them will decay due to the dynamics of such perturbations by the end of the assimilation window. In contrast, even if growing errors are reduced by the assimilation system, what remains of them will amplify by the end of the assimilation window.

The ECMWF and the NCEP ensembles were followed, in 1995, by the Canadian ensemble, which was designed to simulate a wider range of error sources, due to initial uncertainties linked to observation errors and data assimilation assumptions, and also due to model uncertainties.

The initial perturbation strategies used by all the other ensembles operational today have been inspired by these three earlier approaches, although their details are not exactly the same and include upgrades and changes introduced throughout the years.

If we consider model uncertainties, the Canadian ensemble was the first one to include their simulation. Following their example, the simulation of model uncertainties was introduced in the ECMWF ensemble in 1999, using a stochastic approach to simulate the effect of model errors linked to the physical parameterization schemes. This was the first time that a stochastic term was introduced in numerical weather prediction.

At present, four main approaches are followed to represent model uncertainties in the operational ensembles:

- A multi-model approach, where different models are used in each ensemble member; models can differ entirely or only in some components; one example is the seasonal forecasts issued by the Copernicus Climate Change Service.
- A perturbed-parameter approach, where all ensemble integrations are made with the same model, but where the parameters that define the settings of the model components change in each member; one example is the Canadian ensemble.
- A perturbed-tendency approach, where stochastic schemes designed to simulate the random model error component are used to simulate the fact that tendencies are known only approximately; one example is the ECMWF Stochastically Perturbed Parametrization Tendency scheme.
- A stochastic backscatter approach that aims to simulate the upscale energy transfer from the scales below the model resolution to the resolved scales.

Table 7.1 lists the key characteristics of the seven most accurate global ensembles operational today that generate global, medium-range weather forecasts on a daily basis in 2018 (the date when this information was most recently updated):

- The Chinese Meteorological Agency (CMA) ensemble simulated initial uncertainties with global BVs, perturbations designed to emulate the analysis cycle; it did not simulate model uncertainties.
- The Brazilian Centro de Previsão de Tempo e Estudos Climáticos (CPTEC) ensemble simulated initial uncertainties with empirical orthogonal functions (EOF)

Table 7.1 Key characteristics of the seven most accurate global, operational, medium-range ensembles

Center	Initial unc. method	Model unc. method	Horizontal resolution (km)	No. Vert Levels (TOA, hPa)	FC length (d)	No. members per ensemble	No. runs per day (UTC)	No. forecasts per day
CMA	BV	no	70	31 (10.0)	10	15	2 (00/12)	30
CPTEC	EOF	no	120	28 (0.1)	15	15	2 (00/12)	30
ECMWF	SV+EDA	YES	18 / 36	91 (0.01)	0–15 / 15/46	51	2 (00/12) up to d15 (2 per week to d46)	102
JMA	LETKF+SV	YES	40	100 (0.01)	11	26	2 (00/12)	52
KMA	ETKF	YES	32	70 (0.1)	12	25	2 (00/12)	50
MSC	EnKF	YES	50	40 (2.0)	16/32	21	2 (00/12)	42
NCEP	ETR	YES	34 / 55	64 (2.7)	0–8 / 8–16	21	4 (00/06/12/18)	84

Ensembles are listed in alphabetic order (column 1): initial uncertainty method (column 2), model uncertainty simulation (Y/N, column 3), truncation and approximate horizontal resolution (column 4), number of vertical levels and top of the atmosphere in hPa (column 5), forecast length in days (column 6), number of members for each run (column 7), number of runs per day (column 8), and number of forecasts issued per day (column 9). Note that the ECMWF ensemble is also run up to 6 days at 06 and 18 UTC.

computed over a tropical band; it did not simulate model uncertainties.

- The ECMWF ensemble simulated initial uncertainties with a combination of SVs and EDA-based perturbations; it simulated model uncertainties.
- The Japan Meteorological Administration (JMA) ensemble simulated initial uncertainties with a combination of SVs and perturbations generated using a Local Ensemble Transform Kalman Filter (LETKF); it simulated model uncertainties.
- The Korean Meteorological Administration (KMA) ensemble simulated initial uncertainties with an Ensemble Transformed Kalman Filter method with localization; it simulated model uncertainties.
- The Meteorological Service of Canada (MSC) ensemble simulated initial uncertainties with an Ensemble Kalman Filter (EnKF); it simulated model uncertainties.
- The U.S. National Centers for Environmental Prediction (NCEP) ensemble simulated initial uncertainties with perturbations generated with an Ensemble Transform with Rescaling (ETR); it simulated model uncertainties.

The first two columns of Table 7.1 show that different strategies have been followed to generate the initial conditions, and that not all ensembles include a scheme to simulate model error. It also shows the size, resolution, and forecast length of the seven ensembles: size ranges from 15 to 51 members, horizontal resolution from 18 to 120 km, the number of vertical levels from 31 to 137, and the forecast length from 10 to 46 days. These characteristics have been set partly because of theoretical reasons that guided the different groups to develop different perturbation techniques, and partly due to computer power availability. The last column of Table 7.1 lists the total number of ensemble forecasts issued by each center per day (in total, 390 global forecasts were generated every day in 2018), and the relative computing cost of the whole ensemble,

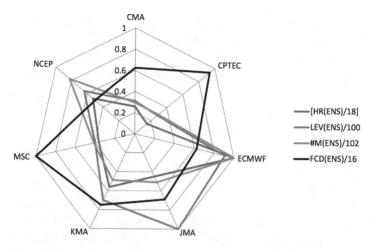

Figure 7.1. Comparison of four key characteristics of the seven global ensembles in 2018: horizontal resolution (HR), number of vertical levels (LEV), number of perturbed members per day (#M), and forecast length (FCD, in days). Values for each driver have been normalized with respect to the costliest value (18 km for HR, 100 for LEV, 102 for #M, and 16 for FCD).

with respect to the ECMWF ensemble (the most expensive computers at that time).

Figure 7.1 shows the difference between the seven global ensembles in four categories: the resolution, normalized by the highest horizontal resolution (18 km, from ECMWF), the maximum number of vertical levels (normalized by 100 levels, used by JMA), the forecast length spanned every day (normalized by 16 days, used by MSC), and the number of ensemble forecasts produced daily (normalized by 102, used by ECMWF). The ensembles with the lines closest to the outside perimeter are the ones with the finer resolution, longer forecast length, and larger membership.

7.2 What is a probabilistic forecast?

A probabilistic forecast indicates how likely an event is to occur. The probability of an event is a number between 0 and 1 (i.e., 100%), where 0 indicates impossibility and 1 certainty.

If we had access to a single, deterministic forecast, we could only state whether an event would or would not occur. We could look at a surface temperature forecast, read the temperature predicted at specific locations, say T_A, and then deduce whether there is a 0% or a 100% probability that the temperature at that location is above, or below, a threshold of interest, T_{THR}. If we took into consideration the surface temperature forecast error statistics, computed by looking at how accurate the past forecasts were, we could add a confidence range around this single forecast. If this confidence range is represented by a typical, bell-shape distribution described by a Gaussian function with a standard deviation defined by the average root-mean-square error of a temperature forecast at that location, E_A, we could transform this single forecast into a probabilistic forecast, $P_{DET}(T)=N_{DET}(T_A,E_A)$, defined by a Gaussian function with mean T_A and standard deviation E_A. If we now computed the probability that the temperature is above or below T_{THR}, we would get a number that is between 0% and 100%. Note that, since the standard deviation of the Gaussian distribution is fixed, the forecast confidence range would be the same every day, and would not reflect the fact that forecast errors are flow dependent.

Instead, if we had access to an ensemble of forecasts $T_{A,j}$, with $j = 1,N$, we could compute the most likely temperature, for example, defined by the ensemble mean, and the probability that the temperature is above T_{THR}, $P_{ENS}(T)$, by counting how many members predict a temperature above or below T_{THR}. Since the probability is computed using independent forecasts, the probability distribution function would not necessarily be a Gaussian function, but it could be of any shape—a shape that depends on the ensemble of forecasts and on the atmospheric flow of the day.

The key difference between the two probability distributions is that while the first probability has a defined form and a fixed standard deviation, and thus does not vary with the atmospheric conditions, the ensemble-based probability depends on the atmospheric conditions. If the ensemble has been properly constructed and it provides accurate and reliable temperature forecasts, then the range spanned by the probability function can be used as a predictor of the forecast skill and to provide a flow-dependent confidence interval around the most likely forecast value.

An example of two ensemble forecasts for temperature at a specific location, based on a 31-member ensemble, is shown in Figure 7.2 to illustrate how one can compute a probabilistic forecast starting from ensembles. Figure 7.2 also shows two

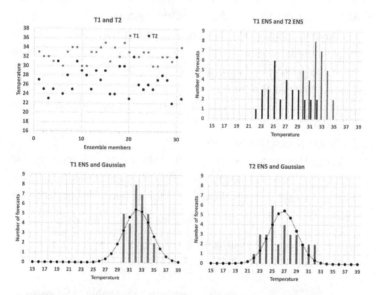

Figure 7.2. Example of two ensemble-based and single-forecast-based probabilistic forecasts. The top-left panel shows two 31-member ensemble temperature forecasts, valid for two days, T1 (squares) and T2 (circles): the x-axis indicates each ensemble member, and the y-axis temperature. The top-right panel shows these two forecasts binned into temperature 1-degree intervals: for each temperature interval [e.g., (24.5–25.5)], each bar shows how many ensemble members predicted a value within that range. The bottom panels show the two 31-member ensemble forecasts, and a Gaussian forecast (black dotted line), which has a mean value equal to the mean of the 31 ensembles, and a standard deviation equal to a climatological-average root-mean-square error (thus, the two Gaussian functions have the same width, but are centered on the ensemble-mean forecast; see text for more details).

Gaussian functions based on a single forecast to illustrate the fact that an ensemble provides a flow-dependent confidence estimate, while a single forecast does not.

The top-left panel shows the two forecasts: for each of the two cases, case 1 (squares) and case 2 (circles), 31 ensemble forecasts are shown. The top-right panel shows the same two forecasts binned into 1-degree intervals $[T-05, T+0.5]$: in other words, each bar shows how many ensemble forecasts predicted a temperature between $[T-05, T+0.5]$, with T going from 15°C to 39°C degrees. From the binned ensembles shown in the top-right panel of Figure 7.2 we can compute probabilistic temperature predictions: if we wanted to compute the probability that the temperature would be above a threshold T_{THR}, we have to count how many members populate all bins above T_{THR}. If we consider, for example, $T_{THR} = 30°C$, the ensemble forecasts for case 1 would give $Prob_1(T>30°C)=71\%$, and for case 2 $Prob_2(T>30°C)=6\%$ (see also Table 7.2). Similarly, if we wanted to know the probability that the temperature would be below 25°C, we would get $Prob_1(T<25°C)=0\%$ for case 1, and $Prob_2(T>30°C)=23\%$ for case 2.

From Figure 7.2 we can see that the range spanned by the ensemble (also called the ensemble spread) is wider in case 2 than in case 1, as indicated by the difference between the

Table 7.2 Key statistics of the ensemble forecasts shown in Figure 7.2

	Case 1		Case 2	
	Ensemble	Gaussian	Ensemble	Gaussian
Ensemble mean <T>	32.3	32.3	26.8	26.8
Minimum value	30.0		22.0	
Maximum value	35.0		32.0	
Interquartile range	2.0		4.0	
Standard deviation	1.49	2.26	2.83	2.26
Prob(T>30°C)	71%	64%	6%	2%
Prob(T<25°C)	0%	23%	0%	15%

maximum and the minimum forecast: 5°C degrees in case 1 and 10°C degrees in case 2. Another possible indicator of the range spanned by an ensemble is the interquartile range, which is 2°C in case 1 and 4°C in case 2.

Assuming that the ensemble is reliable, the different spread implies that there is a relatively smaller forecast uncertainty in case 1 than in case 2, and thus that case 1 is more predictable than case 2. If instead of predicting a full probability distribution function we only wanted to know what is the most likely forecast, we could use the ensemble mean: in case 1, $<T_1>=$ 32.3°C, and in case 2, $<T_2>= 26.8°C$.

The two bottom panels of Figure 7.2 include also a probabilistic forecast defined by a Gaussian distribution with the mean value equal to the ensemble-mean forecast, and with a standard deviation equal to the average climatological temperature error. Table 7.2 also reports these values. Since the two Gaussian functions have been centered on the ensemble-mean forecasts, if we were interested only in predicting the most likely temperature, there would not be any difference from the forecast issued by the ensembles. But if we wanted to assess the forecast confidence using the climatological spread, we would get the same value in the two cases, and thus we would not be able to say whether one situation was easier or more difficult to predict than the other. The two Gaussians could also be used to compute probabilities, but again care should be taken in using these "Gaussian type" probability functions, since in reality the range of possible forecast states could follow a different distribution. This can be seen also in Figure 7.2, which shows that the ensemble distributions do not necessarily look like a Gaussian. For example, in case 2 the ensemble forecasts give an indication of two possible weather scenarios: one characterized by a temperature below 26°C, and one with temperature between 27°C and 32°C. By contrast, the Gaussian forecast predicts a unique scenario with temperature between 25°C and 29°C.

7.3 How can we communicate forecast uncertainty?

Forecast uncertainty can be communicated with a probabilistic language, by issuing forecast products that display a full range of possible values with their probability.

Using a deterministic language gives the wrong impression that weather forecasts are not affected by uncertainty, although we know that this is not the case. Deterministic maps and diagrams should be replaced by maps or diagrams that show the level of uncertainty and the full range of possible outcomes.

Figures 7.3 to 7.10 show examples of ECMWF weather forecasts issued on a (randomly chosen) day, August 5, 2022: they should give a good overview of how forecast uncertainty can be communicated, either in terms of probabilities, in terms of an index computed from the probability distribution function of forecast states, or as an ensemble-gram that shows the forecast range spanned by the ensemble at specific locations.

Let us start by looking at traditional, single forecasts. Figure 7.3 shows a 96-hour forecast of total precipitation and mean sea-level pressure (top panel) and of 2-meter temperature and 30-meter wind (bottom panel), valid for August 9, 2022, at 00UTC. The forecasts, produced by the ECMWF single high-resolution model, provide information on the most likely scenario, while they do not provide any indication of the forecast level of confidence.

Let us now look at some ensemble-based forecasts. Figure 7.4 shows two 96-hour ensemble-based probabilistic forecasts: the probability that the accumulated precipitation is less than 0.1 mm/d between forecast day 5 and 10 (dry conditions; top panel), and the probability that the 2-meter temperature is above 30 degrees (bottom panel). The first field indicates that there are areas that will almost certainly be dry, and others where there is more uncertainty and the probabilities are far from the two extremes (0% and 100%). Similar information is contained in the second forecast, for the 2-meter temperature.

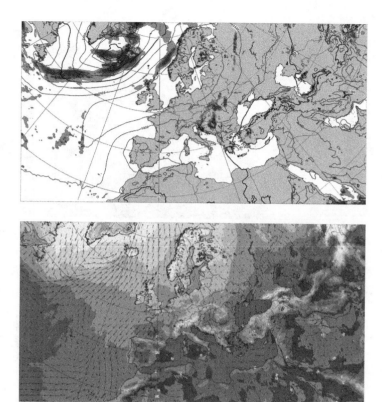

Figure 7.3. Example of two single forecasts issued on August 5, 2022, at 00 UTC. The maps show the forecast of total precipitation and mean sea-level pressure (top panel), and of 2 m temperature and 30 m wind (bottom panel), valid for August 9, 2022, at 00UTC (+96h forecast). Precipitation shading has a contour interval for 0.5, 2, 4, 10, 25, and 50 mm/d; mean sea-level pressure has one isoline every 5 hPa; 2 mT has a different shading color every 4 degrees, and 30 m wind is shown in knots (equivalent to 0.514 m/s). (Source: ECMWF)

Figure 7.5 shows another type of ensemble-based forecast, defined by comparing the probability distribution function of forecast states and the climatological distribution. More precisely, the top panel shows the Extreme Forecast Index (EFI) for 2-meter temperature. The EFI varies between 0 and 1, and it indicates how close or different the forecast distribution is from climatology: areas with the largest values are the ones

Thursday 04 August 2022 1200 UTC ECMWF t+120-240 VT: Sunday 14 August 2022 1200 UTC
Surface: Total precipitation of less than 0.1mm

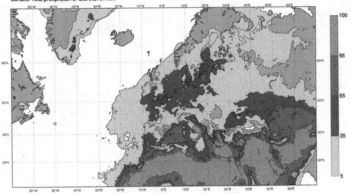

Probabilities: maximum 2 m temperature, last 6 hours

Base time: Fri 05 Aug 2022 00 UTC Valid time: Tue 09 Aug 2022 00 UTC (+96h) Area : Europe Event threshold : >30 C

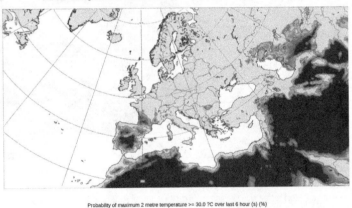

Probability of maximum 2 metre temperature >= 30.0 ?C over last 6 hour (s) (%)

Figure 7.4. Example of two ensemble-based probabilistic forecasts issued on August 5, 2022, at 00 UTC. The maps show the probability that the accumulated precipitation is less than 0.1 mm/d between forecast day 5 to 10 (dry conditions; top panel), and the probability that the 2 m temperature is above 30 degrees (bottom panel) valid for August 9, 2022, at 00UTC (+96h forecast). Precipitation probability has shading contours at 5-35-65-95-100%; 2 m temperature probability has shading contours at 5-10-25-50-75-90-95-100%. (Source: ECMWF)

Fri 05 Aug 2022 00UTC ©ECMWF t+72-96h VT: Mon 08 Aug 2022 00UTC - Tue 09 Aug 2022 00UTC
Extreme forecast index and Shift of Tails (black contours 0,1,2,5,8) for 2m mean temperature

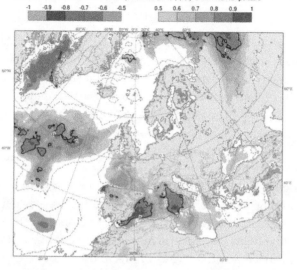

Thu 04 Aug 2022 00UTC ©ECMWF VT: Mon 08 Aug 2022 00UTC - Tue 09 Aug 2022 00UTC 72-96h
2m mean temperature (in °C) Model climate Q90 (one in 10 occasions realises more than value shown)

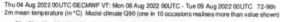

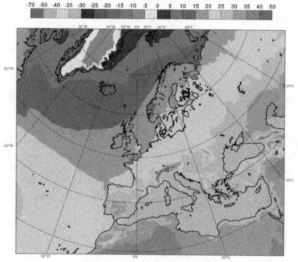

Figure 7.5. Example of an ensemble-based forecast issued on August 5, 2022, at 00 UTC. The top panel shows the Extreme Forecast Index (EFI) for 2 m temperature valid for August 9, 2022, at 00UTC (+96 hour forecast). The EFI is computed by comparing the probability distribution function of forecast states, with the climatological distribution of past states: it gives a value, between –1 and 1, of how extremely cold, normal, or extremely hot the forecast situation is compared to climatology. The bottom panel shows a reference "distribution" map: it shows the 90th percentile of 2 m temperature climatology computed by considering the past 20 years. The EFI shading contouring goes with 0.1 steps and shows only values between –1 and –05, and 05 and 1; the 2mT 90th percentile map has shading contours every 5 or 10 degrees, from –70 to +50 degrees. (Source: ECMWF)

where extreme weather conditions are predicted. The bottom panel shows a reference field that can be used to understand how high the extreme values could be: it shows the 90th percentile of 2-meter temperature climatology computed by considering the past 20 years. The EFI depends on the distribution of forecast states: if the uncertainty is very large, the EFI tends to be small. Regions with high EFI values are the ones where extreme conditions can occur.

Figure 7.6 is similar to Figure 7.5, but it shows in one map the EFI for three surface weather variables: 2-meter temperature, 10-meter wind speed, and total precipitation. Thus, this map highlights the regions where extreme weather conditions linked to temperature, wind, or precipitation could occur.

Now let us consider a very specific weather event, a tropical storm, for which it is very important to predict both its path and intensity. Figure 7.7 shows two examples of ensemble-based tropical-storm strike-probability forecasts, valid for the next 10 days, issued for Hurricane Sandy, which hit New York

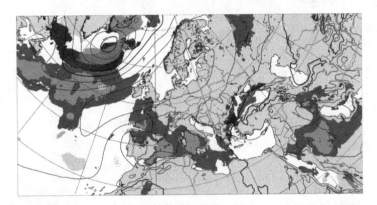

Figure 7.6. Example of an ensemble-based forecasts issued on August 5, 2022, at 00 UTC. The map shows the Extreme Forecast Index (EFI) for three surface weather parameters valid for August 9, 2022, at 00UTC (+96-hour forecast): the EFI for 2 m temperature is shown with shading (light gray for cold extremes and dark gray for hot extremes); the EFI for 10 m wind speed is shown with medium gray shading; the EFI for total precipitation is shown with black shading. The contours show the ensemble-mean mean-sea-level-pressure forecast. (Source: ECMWF)

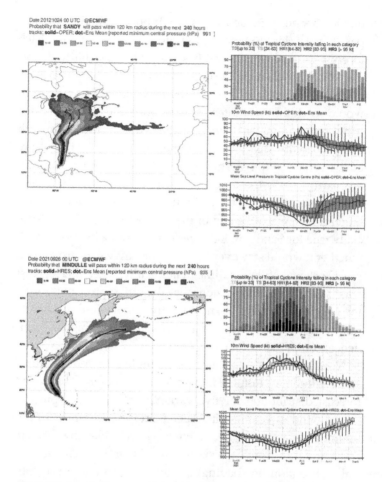

Figure 7.7. Example of two ensemble-based hurricane strike-probability maps. The top panel shows the forecast of Hurricane Sandy, which hit New York on October 29, 2012, issued on October 24, 2012, at 00UTC. The bottom panel shows the forecast of Super Typhoon Mindulle, which passed very close to Japan, issued on September 26, 2021, at 00UTC. Each panel shows the strike probability, that is, the probability that the tropical storm would pass within a 120 km radius, the probability that the storm would reach different intensity categories, the probability distribution of 10 m wind near the center of the storm, and the probability distribution of the mean sea-level pressure of the center of the storm. (Source: ECMWF)

on October 29, 2012, and for Super Typhoon Mindulle, which passed very close to Japan at the end of September 2021. Each forecast shows four probabilities: the strike probability, that is, the probability that the tropical storm would pass within a 120 km radius of each grid point; the probability that the storm would reach different intensity categories; the probability distribution of 10-meter wind near the center of the storm; and the probability distribution of the mean sea-level pressure of the center of the storm. Note that the strike probabilities are narrower for the first few forecast days, when the forecast uncertainty is small, while they spread out and cover a larger area toward the end of the forecast range. This is especially true for Sandy: in fact, predicting 1 week before landfall whether it would stay over the ocean or turn westward and hit New York was still very difficult at the time when this forecast was issued. By contrast, note that the strike probability map for Mindulle remains narrow for the whole 10-day forecast range, indicating less forecast uncertainty (i.e., higher confidence) compared to Sandy.

Now let us consider another example of a forecast for a specific location, Rome. Figure 7.8 shows an ensemble-based forecast distribution for five variables: cloud cover, precipitation, wind direction, wind speed, and maximum and minimum 2-meter temperature. For each variable, the forecast distribution (box-and-whiskers symbols) and the climatological distribution (in shading) are overlaid, so that a user can compare the two, judge how extreme the weather condition could be compared to climatology, and gauge the forecast confidence level. Consider, for example, the daily maximum and minimum temperature: the forecast indicates that for the first 5 days both values are going to be very close to the maximum values observed in the past 20 years, and that the uncertainty in this forecast is very small.

Finally, let us now move to a longer forecast range, to seasonal prediction. Figure 7.9 shows an ensemble-based probabilistic tercile forecast of the 2-meter temperature distribution

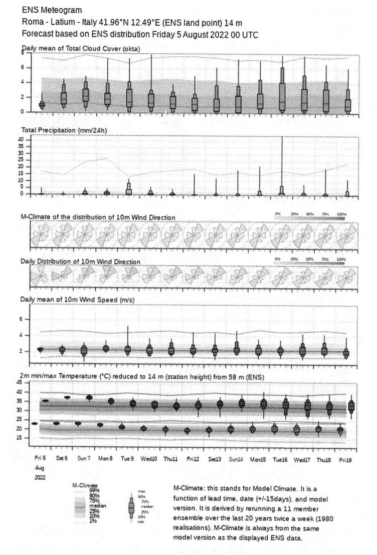

ENS Meteogram
Roma - Latium - Italy 41.96°N 12.49°E (ENS land point) 14 m
Forecast based on ENS distribution Friday 5 August 2022 00 UTC

Daily mean of Total Cloud Cover (okta)

Total Precipitation (mm/24h)

M-Climate of the distribution of 10m Wind Direction

Daily Distribution of 10m Wind Direction

Daily mean of 10m Wind Speed (m/s)

2m min/max Temperature (°C) reduced to 14 m (station height) from 58 m (ENS)

Fri 5 Sat 6 Sun 7 Mon 8 Tue 9 Wed10 Thu11 Fri12 Sat13 Sun14 Mon15 Tue16 Wed17 Thu18 Fri19
Aug
2022

M-Climate: this stands for Model Climate. It is a function of lead time, date (+/-15days), and model version. It is derived by rerunning a 11 member ensemble over the last 20 years twice a week (1980 realisations). M-Climate is always from the same model version as the displayed ENS data.

Figure 7.8. Example of an ensemble-based forecast at a specific location, Roma, issued on August 5, 2022, at 00 UTC and valid for the next 15 days. The "ensemble-gram" shows the distribution of possible forecasts for five variables: cloud cover (top), precipitation, wind direction, wind speed, and maximum and minimum 2 m temperature (bottom). For each variable, the figure shows the forecast distribution (box-and-whiskers symbols) and the climatological distribution (in shading). (Source: ECMWF)

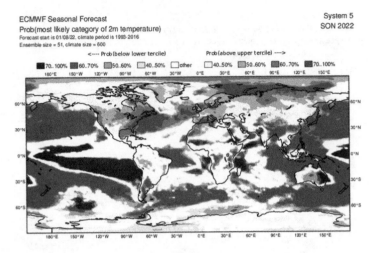

Figure 7.9. An example of an ensemble-based probabilistic tercile forecast of the 2 m temperature distribution, issued on August 1, 2022, and valid for September-October-November (2-to-5-month forecast). The map shows the probability of the most likely 2 m temperature tercile category (upper tercile, middle tercile, and lower tercile). The probability contours are in percentage categories, black for the lower tercile and dark gray for the upper tercile. (Source: ECMWF)

valid for the forthcoming months. The map shows, for each grid point, the most likely 2-meter temperature tercile category (upper tercile, middle tercile, and lower tercile) and its probability. Dark/light areas identify regions where the predicted temperature is in the lower/upper tercile. White areas are the ones where climatological conditions are predicted. The intensity of the colors can be used as an indication of the forecast confidence level.

Figure 7.10 shows the ensemble-based 2-meter temperature distribution forecast for the monthly average temperature over Southern Europe valid for the next 6 months. The "ensemble-gram" shows the forecast probability distribution, represented by the minimum and maximum, 25th and 75th quantile, and median, and the climatological distribution computed for the past 20 years. The width of the forecast symbols gives an indication of the forecast level of uncertainty. The comparison

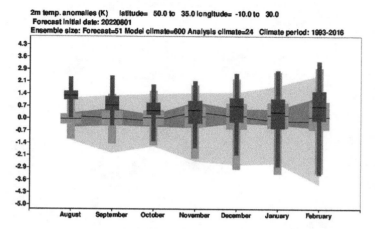

2m temp. anomalies (K)   latitude= 50.0 to 35.0 longitude= -10.0 to 30.0
Forecast initial date: 20220801
Ensemble size: Forecast=51 Model climate=600 Analysis climate=24 Climate period: 1993-2016

Figure 7.10. An example of an ensemble-based 2-m temperature distribution forecast for the average temperature over Southern Europe issued on August 1, 2022, and valid for the next 6 months. The "ensemble-gram" shows the forecast probability distribution, represented by the minimum and maximum, 25th and 75th quartile, and median, and the climatological distribution computed for the past 20 years. (Source: ECMWF)

between the width of the forecast and the climatological distribution can be used to assess the forecast confidence level, and the separation between the forecast and climatological distributions gives an indication of how anomalous the next months could be.

Figure 7.11 shows two ensemble-based predictions of the anomaly (i.e., the difference between the forecast value and climatology) of the average sea surface temperature over an area in the central tropical Pacific, the so-called El Niño 3.4 area, valid for the next 6 months. Each figure shows 51 ensemble members and the observed values. The average sea surface temperature in the El Niño 3.4 area is used to characterize the ocean and atmosphere weather conditions in the Pacific: large positive (large negative) sea surface temperature anomalies characterize warm El Niño (cold La Niña) conditions. The two ensemble forecasts can be used to estimate the trend and gauge the forecast confidence.

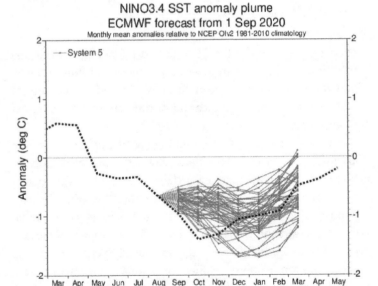

Figure 7.11. Two examples of ensemble-based predictions of the anomaly of the sea surface temperature averaged within an area in the central tropical Pacific (the El Niño 3.4 area), issued on September 1, 2019 (top panel) and 2020 (bottom panel), valid for the next 6 months. Each figure shows the 51 ensemble members (gray lines) and the observed values (dotted line). (Source: ECMWF)

These are just a few of the ensemble-based probabilistic forecast products that are issued daily to communicate forecast uncertainty: they should have illustrated the value of ensembles.

7.4 How can we make decisions using probabilistic forecasts?

Consider a user that has to decide whether to incur a cost C to protect against a loss L (with C<L) if a weather event occurs. Using a probabilistic weather forecast, the user should decide to take the action if the probability of the weather event is higher than an optimum threshold. If the probabilistic forecast is reliable, the optimum threshold is C/L.

Consider a decision maker interested in protecting from the impact of a possible weather event, for example, precipitation above a certain threshold or a heat wave. Suppose that if the weather event occurs, he will incur a loss L, but that he can avoid this loss if he takes an action now that involves a cost C (with $C<L$). Table 7.3 summarizes the possible outcomes linked to this simple cost-loss model.

This simple model can be used to show that, if the decision maker has access to a reliable probabilistic forecast, his best strategy is to act if the forecast probability is higher than C/L.

Suppose that a user will incur a loss of 1,000 EUR if it rains more than 5 mm over 24 hours, but that he can decide to pay 100 EUR now to protect against this loss. Suppose that this user

Table 7.3 Simple cost/loss model, for a user who can take an action that has a cost C, to protect against a potential loss L linked to the occurrence of an event (with $C<L$)

Cost/loss matrix		The event occurs	
		Yes	No
A protection action is taken	Yes	C	C
	No	L	0

is given access to a single deterministic and a reliable probabilistic forecast:

- *Single forecast*: this forecast would tell him whether it will or will not rain more than 5 mm/d. The user will decide to take the protection action if rain is predicted to be above 5 mm/d.
- *Probabilistic forecast*: this forecast will give the probability that rain could be higher than 5 mm/d, $PR(TP>5)$. In this case, given that its C/L ratio is 0.1, the user should take the protection action if $PR(TP>5)>0.1$.

There is a forecast verification metric called the potential economic value (PEV) of a forecast, which is based on this decision model. It measures the savings linked to using a forecast compared to the savings linked to having access to a perfect forecast, considering users with a whole range of C/L from 0 to 1. PEV can be applied both to a single and a probabilistic forecast, and it is routinely computed at the major operation centers. PEV statistics based on a large number of cases are used to assess the forecast potential economic value.

As an example, Figure 7.12 shows the PEV of the ECMWF operational 6-day forecast of precipitation in excess of 1 and 5 mm/d over Europe and North Africa, computed considering all forecasts issued from April 1 to June 30, 2022. Each panel shows two curves: the PEV or the ECMWF single high-resolution forecasts (solid lines), and the PEV of the ECMWF ensemble-based probabilistic forecasts (dashed lines). The fact that the dashed curves lay above the solid curves means that the ensemble PEV is higher. Note that the precise value of the PEV depends on the user's cost/loss ratio, and that for some users neither the single high-resolution nor the ensemble forecasts have a PEV higher than zero (i.e., they do not beat a climatological forecast, since a zero PEV means that the forecast is equivalent to climatology).

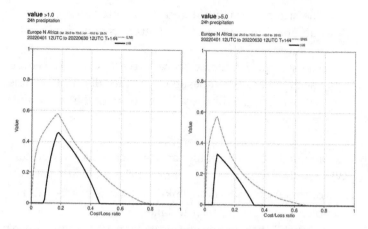

Figure 7.12. Potential economic value (PEV) of ECMWF operational 6-day forecasts of precipitation in excess of 1 mm/d (left panel) and 5 mm/d (right panel). PEV has been computed considering forecasts valid over Europe and North Africa at available station locations, for the period April 1 to June 30, 2022. Each panel shows the PEV of ECMWF operational single high-resolution forecasts (solid lines) and ensemble-based probabilistic forecasts. (Source: ECMWF)

7.5 What is a scenario forecast?

A scenario forecast provides the probability that different weather events could occur, with the probabilities estimated using a priori information (e.g., based on climatology) and probabilistic forecasts.

A scenario forecast answers the following user questions: What weather patterns might occur in the future? What is the probability that a range of weather patterns could occur? These questions can be answered using ensemble forecasts. Some of the products shown above can be considered as scenario forecasts:

- The probabilistic forecasts of rainfall less than 0.1 mm/d (Figure 7.4) answer the question "whether a dry or wet weather scenario might occur."
- The EFI forecasts (Figures 7.5 and 7.6) answer the question "whether and where extreme weather conditions could occur."

- The strike probability forecasts (Figure 7.7) answer the question "whether a tropical storm might hit a certain region or stay offshore."
- The sea surface temperature forecasts over the central Pacific answer the question "whether the next months will be characterized by El Niño or La Niña weather conditions."

Compared to the traditional weather forecast question "What is the weather going to be like?" a scenario forecast can address users' questions linked to events of interest. For example, a user might need to take action if there is a very small probability that a specific weather event can cause major damages (e.g., linked to a rough state of the sea that could make it impossible for a shift to enter a port). With an ensemble of forecasts, we can identify which weather patterns could occur, and compute their probability of occurrence.

7.6 What is a cluster analysis?

A cluster analysis is a way to condense all the information contained into an ensemble into only a few, relevant weather patterns.

Analyzing an ensemble of tens of forecast maps and taking decisions based on them can be very daunting. One way to simplify the process is to summarize the ensembles into a few, key patterns, by identifying the ensemble members that are more similar to each other and grouping them together into a cluster. Then, each cluster could be identified by a "representative member," to which we could assign a probability of occurrence linked to how many members of the ensemble are part of the cluster. A representative member could be defined as the cluster member closest to all the cluster's members and furthest away from the members of the other clusters.

Although different methods can be used to define the clusters, they all use an objective metric to measure the similarity of two fields. One of the key parameters that determine the outcome of a cluster analysis is how many clusters we want to identify. If N forecasts are grouped into a few (say, two or three) clusters, then a lot of detailed information that the different ensemble members contain is lost. By contrast, if they are grouped into a slightly large number of clusters (say, five or six), more detailed information could be kept. Most of the operational cluster analysis limits the maximum number of clusters to six, which provides a good balance between the need to condense the ensemble information and to avoid losing potentially valuable details. In this way, predictable cases, characterized by a small divergence among the ensemble members, would on average be characterized by a small number (say, one to three) of clusters, while less predictable cases would be characterized by a large number (say, four to six) of clusters.

Figure 7.13 shows an example of a cluster analysis applied to the ECMWF ensembles. The four clusters have been computed on the 51 ensemble forecast trajectories that span the day 5-to-7 forecast range, over Europe. The ECMWF clustering algorithm enforces a maximum of six clusters, while there is no minimum number of clusters. The three columns show the four clusters' representative members at forecast 5, 6, and 7. Each map shows the 500 hPa geopotential height field, which describes the atmospheric flow at about a 5,000 meter height, and its anomaly with respect to the climatological field, that is, the difference between the forecast and the average 500 hPa field for that day of the year computed considering the past 20 years. The number of clusters' members can be used to assign a probability of occurrence to each cluster: in this case, the clusters' population (21, 17, 10, 3 members out of 51) can be translated into the clusters' probability of occurrences (41%, 33%, 20%, 6%).

Cluster scenario

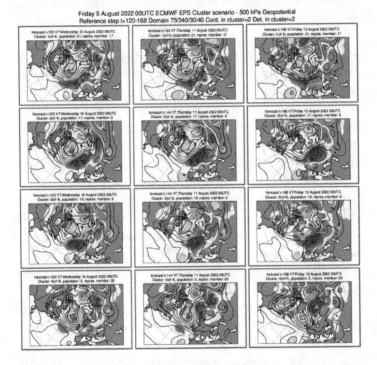

Figure 7.13. An example of cluster analysis applied to ensemble forecasts issued on August 5, 2022, at 00UTC. The four clusters have been computed over a 3-day forecast trajectory that includes forecast days 5, 6, and 7. The left column shows the four clusters' representative members valid for forecast day 5 (+120h), starting from cluster 1 (21 members), cluster 2 (17 members), cluster 3 (10 members), and cluster 4 (3 members). The other two columns show the same representative members at forecast days 6 and 7. Each panel shows the 500 hPa geopotential field and its anomaly with respect to climatology (in shading). (Source: ECMWF)

7.7 How do we measure the accuracy and reliability of a probabilistic forecast?

The accuracy and reliability of probabilistic forecasts are measured using a range of metrics that are sensitive to the characteristics of the whole probability distribution function of forecasts states, computed considering a large number of cases. These metrics are the equivalent, for ensemble forecasts, of the ones used to assess the accuracy and skill of single forecasts.

Denote with $p_k(x_j,t)$ the probability that event k, with $k = 1,r$ would occur at grid point $x_j \in \Sigma_G$, and $o_k(x_j,t)$ the corresponding probability that event k would occur, defined by a very narrow distribution with a peak at the observation (or analysis) value, and with a width that depends on the observation (or analysis) uncertainty. Thus, $p_k(x_j,t)$ and $o_k(x_j,t)$ with $k = 1,r$ spanning all possible outcomes describe the probability distribution functions of forecast and observation states.

The equivalent of the root-mean-square error used to compute the accuracy of single forecasts is the ranked probability score for the probability density forecast (PDF). If we consider all the grid points N_G within the area Σ_G of interest, the RPS is defined in the following way:

$$RPS(p,v;t) = \sqrt[2]{\frac{1}{N_G} \sum_{N_G} \left[RPS(x_j,t) \right]^2}. \tag{7.1}$$

here $RPS(x_j,t)$ is the ranked probability score at the point x_j:

$$RPS(x_j,t)^2 = \frac{1}{r-1} \sum_{i=1}^{r} \left[\sum_{k=1}^{i} p_k(x_j,t) - \sum_{k=1}^{i} o_k(x_j,t) \right]^2. \tag{7.2}$$

it was the case for the metrics used to evaluate a single forecast, if the area is large and includes points that have very different latitudes, a weight that depends on the latitude of the grid point x_j should be included in the computation, to take into account the fact that the area that each point represents changes as we move from the equator to the poles.

Two other metrics that focus on some particular events (i.e., that consider only one of all the r possible outcomes) can also be applied to assess the accuracy and reliability of this type of probabilistic forecast are Brier score and the area under a relative operating characteristic curve (ROC).

For each metric, as for single forecasts, we can define skill scores by computing the relative score with respect to a reference forecast, which is usually defined by climatology (i.e., a PDF forecast defined by the statistics of the weather observed in the past 20 years).

7.8 What are reforecasts and reanalyses?

A reforecast is a forecast of a past case, reproduced with a more recent numerical weather prediction system. A reanalysis is an analysis of a past event, reproduced by assimilating all available data using a more recent data assimilation. Reforecasts and reanalyses are used to assess forecast quality and to monitor the evolution of the Earth's climate.

Reforecasts are generated to assess the performance of the most recent prediction systems, for example, the operational version, in some specific and very interesting cases that occurred in the past, such as cases that have caused major damages or that were difficult to predict with the operational model version of the time. Reforecasts are also required if we want to compute the model climatology, which is needed to generate some calibrated products.

Reanalyses are generated to test the performance of the most recent data assimilations on past cases, and/or to generate a more accurate estimate of the state of the atmosphere of past cases. Since reanalyses usually cover many decades, their computing cost is high, and thus they are generated only every 7–10 years. Compared to the operational analyses of the time they re-process, reanalyses provide a better estimate of the past state of the atmosphere since they use a more recent, state-of-the-art model and data assimilation. Furthermore, sometimes reanalyses can be produced by assimilating observations that, at the time, might have arrived too late to be included in the assimilation process.

Since reanalyses cover many decades, they provide a very long time record of consistent analyses, that is, estimates of the state of the Earth system generated using the same model, the same resolution, and the same data assimilation. By contrast, the operational analyses show discontinuities since they are produced with the operational model versions that change usually about once a year. Sometimes reanalyses also include reprocessed observations, that is, observations regenerated from the raw data by using more recent observation operators that transformed the raw data into data that could be assimilated, and this could also lead to improved estimates of the state of the Earth system.

Reforecasts are cheaper to produce than reanalyses; thus, they can be produced daily, in parallel with the operational forecasts, so that calibrated forecast products that require an estimate of the (operational) model climate can be produced. Examples of these products are the Extreme Forecast Index, which is computed by comparing the forecast probability distribution function of forecast states with the model climatological distribution function, or ensemble-grams that display the latest forecast distribution and the model climatological distribution.

By comparing reforecasts with the operational forecast of the time, operational prediction centers can also monitor progress, assess whether years of work and investments have led to advances, and investigate whether problems linked to model deficiencies that affected past operational forecasts have been solved. For example, they have been used to assess whether changes in resolution or in moist processes have led to improvements in the prediction of tropical storms, or whether the introduction of a coupled three-dimensional dynamical ocean model has led to improvement in the prediction of large-scale, low-frequency phenomena over the extra-tropics.

"Climate reanalyses" are reanalyses generated to study climate trends. They usually cover at least 50 years, and some of them extend even beyond 100 years, and they are generated in a way to limit the impact of changes in the observing system on the estimates of the state of the Earth system. In fact, care should be taken in selecting which observations to include in the production of climate reanalyses. Since over time the number and type of the observation can change dramatically (think, e.g., of the changes between the 1960s–1970s and the 1980s–1990s, linked to the increasing use of satellite data), this can affect the characteristics of the reanalyses. For example, having many more observations can improve substantially the initialization of the small scales, and this can leave a signature in the intensity of certain features in the reanalyses that could be attributed wrongly to climate change, although it is due to changes in the observing system. Because of this, climate reanalyses are generated with a more conservative set of observations that do not include, for example, satellite data, to avoid drastic variations in the number and quality of the assimilated observations.

There are only a very few major centers that have produced good-quality global reanalyses: the Japan Meteorological Agency (JMA; JRA-55 is their latest reanalysis that spans the period from 1958 to date), the National Aeronautics and Space Administration (NASA; MERRA is their latest reanalysis that spans the period from 1979 to date), ECMWF (ERA5 is their latest reanalysis that spans the period from 1950 to date), and the National Oceanic and Atmospheric Administration (NOAA; their v2 reanalysis spans the period from 1979 to date). Some of these centers have also generated climate reanalyses (e.g., ERACLIM and ERACLIM2 by ECMWF, and 20CR by NOAA).

There are now also regional reanalyses generated by nesting limited-area models into global reanalyses. Some of them have also reassimilated local observations to generate higher-resolution, local reanalyses. Usually, to contain production costs, these regional reanalyses cover a shorter period than global reanalyses, say the last 20 years, and only very few of them are generated by assimilating extra, local observations that had not been assimilated by the "parent," global reanalyses into which they are nested.

7.9 Why are reanalyses and reforecasts useful?

Reanalyses and reforecasts provide a consistent dataset of analyses and forecasts generated using the same model and data assimilation spanning a long period. Reanalyses are extremely valuable to assess climate trends in areas and/or parts of the atmosphere where there are few observations. Reforecasts allow a thorough and statistically robust assessment of forecast performance.

The combined use of reforecasts and reanalyses has led to major improvements in the understanding of climate trends and predictability.

The availability of a consistent dataset of analyses spanning several decades and generated using the same assimilation system (model, methodology, statistical assumption) has allowed us to understand how the Earth's climate has been changing also in areas covered by few observations. Because of this, reanalyses spanning several decades are now part of the operational suite of the most advanced numerical prediction centers.

Reforecasts are used to understand the predictability of rare events, events that occur a few times every year. Consider, for example, El Niño and La Niña, events that occur in the tropical Pacific every 3 to 5 years. Only with reforecasts spanning many years can we accumulate enough forecast error statistics to be able to conclude how far ahead these events can be predicted. The same could be stated of rare flood events or droughts hitting some specific regions.

7.10 Key points discussed in Chapter 7 "Dealing with uncertainties and probabilistic forecasting"

These are the key points discussed in this chapter:

- Although there is not a unique way to generate ensembles, they have all been designed to simulate all relevant sources of forecast errors, linked to initial conditions and model uncertainties.
- A probabilistic forecast indicates how likely an event is to occur or the likelihood of a range of possible forecast scenarios.
- Uncertainty can be communicated with probabilities, by giving to the users the full range of possible future scenarios and their probability of occurrence.
- Clusters can be used to condense the information contained in an ensemble of forecasts into a few, relevant weather scenarios.
- Probabilistic forecasts are assessed over a large number of cases, using a range of metrics based on the comparison of a forecast probability distribution function and observations, or analyses.
- Reforecasts and reanalyses are forecasts and analyses of past cases, reproduced using a more recent model and data assimilations.
- Reforecasts and reanalyses are used to generate operational products, to produce more robust assessments of the quality of numerical weather assimilation and prediction systems.
- Reanalyses, and more specifically climate reanalyses, can be used to monitor climate trends.

8

THE FORECAST
SKILL HORIZON

In this chapter we review the progress in numerical weather prediction, introduce the forecast skill diagram and the concept of the forecast skill horizon, and discuss predictability issues. More specifically, we will be addressing the following questions:

1. Are weather forecasts more accurate and reliable today than in the past?
2. How did we manage to improve the accuracy and reliability of weather forecasts?
3. Can we visualize in a single diagram our prediction capabilities?
4. Why does the forecast skill depend on the phenomena we are trying to predict?
5. Are extreme events more difficult to predict than the "normal" weather?
6. What is the minimum spatial scale that a model can simulate realistically?
7. What is the minimum spatial scale properly resolved in data assimilation?
8. How can we further extend predictability?

8.1 Are weather forecasts more accurate and reliable today than in the past?

Yes, they are. In the medium range (forecasts valid for 3 to 15 days), single forecasts have, on average, improved by about 1 day per decade, and probabilistic forecasts by about 1.5 days per decade. In the extended range (forecasts valid for 2 to 6 weeks), they have improved by about 5 to 7 days every decade. In the long range (forecasts valid for 1 to 12 months), seasonal forecasts that did not exist 30 years ago are now able to provide skillful large-scale patterns months ahead. Since today models use a resolution which is finer than in the past, we are also able to predict scales that were impossible to predict in the past.

Figure 3.7 (see Chapter 3) illustrates in a traditional way the trend in forecast skill for the European Centre for Medium-Range Weather Forecasts (ECMWF) single high-resolution forecast over both the Northern Hemisphere (top, bold lines) and the Southern Hemisphere (bottom, thin lines). In this figure, each curve indicates the anomaly correlation coefficient of the 500 hPa forecasts issued between 1981 and 2022: for example, it shows that in 2015 the 7-day forecasts valid over both hemispheres had an average anomaly correlation coefficient of about 0.75. Curves with different shadings of gray indicate the quality of forecasts valid for different forecast ranges. From Figure 3.7 we can see that the 7-day forecast in 2022 had an anomaly correlation coefficient of about 0.8, similar to the anomaly correlation coefficient of the 5-day forecast issued in 2000, thus indicating a gain in predictability of about 2 days in 22 years.

Figures 8.1 and 8.2 show the trends in forecast accuracy and reliability in a different way, by tracking the forecast lead time when a certain accuracy threshold is crossed, for single and ensemble forecasts.

Figure 8.1 shows the trend in the accuracy and reliability of ECMWF operational forecasts of the large-scale flow, represented by the 500 hPa geopotential height and the 850 hPa temperature over the Northern Hemisphere (i.e.,

Lead time of anomaly correlation coefficient scores (ACC) of 500 hPa height falls to 80%

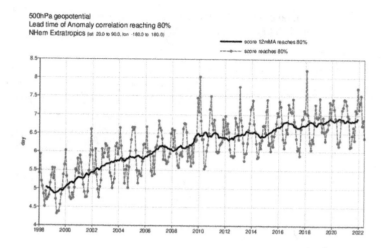

Lead time of continuous ranked probability skill score (CRPSS) of 850 hPa temperature forecasts falls to 25%

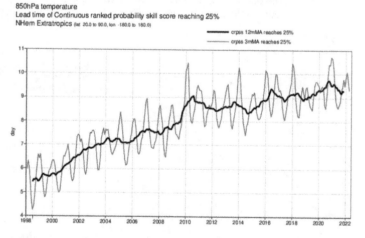

Figure 8.1. Trend in the forecast skill of the 500 hPa geopotential height and the 850 hPa temperature, over the Northern Hemisphere. The top panel shows the forecast time when the anomaly correlation coefficient of ECMWF single high-resolution forecasts of the 500 hPa geopotential falls below 0.8. The bottom panel shows the forecast time when the continuous ranked probability skill score of ECMWF probabilistic forecasts of the 850 hPa temperature falls below 0.25. The gray line in the top panel shows the monthly mean value, and in the bottom panel it shows the 3-month running mean. The black lines show the 12-month running average. Forecasts are verified against analyses. (Source: ECMWF)

Lead time of stable equitable error in probability space (1-SEEPS) of 24 hour precipitation reaching a threshold

total precipitation
Stable Equitable Error in Probability Space
Extratropics (lat -90 to -30.0 and 30.0 to 90, lon -180.0 to 180.0)

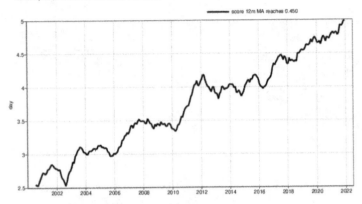

Lead time of continuous ranked probability skill score (CRPSS) of 24 hour precipitation forecasts falls to 10%

total precipitation
Continuous ranked probability skill score
Extratropics (lat -90 to -30.0 and 30.0 to 90, lon -180.0 to 180.0)

Figure 8.2. Trend in the skill of the 24-hour accumulated precipitation over the extra-tropics. The top panel shows the forecast time when (1-SEEPS) falls below the 0.45 value, where SEEPS is the Stable Equitable Error in Probability Space Score of ECMWF single high-resolution forecasts of the 24-hour accumulated precipitation. The bottom panel shows the forecast time when the continuous ranked probability skill score of ECMWF probabilistic forecasts of total precipitation falls below 0.10. The black lines show the 12-month running average. Forecasts have been verified against station observations. (Source: ECMWF)

considering all grid points with a latitude greater than 30°N). The gray line in the top panel shows the monthly mean value, while in the bottom panel it shows the 3-month running mean. The black lines in both panels show the 12-month running average. Forecasts are verified against analyses. The top panel of Figure 8.1 shows the forecast time when the anomaly correlation coefficient of ECMWF single high-resolution forecasts of the 500 hPa geopotential falls below an anomaly correlation coefficient value of 0.8. The figure shows an upward trend: for example, we can see that the 0.8 value was crossed in 2002 at forecast day 5.5 and in 2022 at forecast day 6.9, thus indicating a gain in predictability of 1.4 days over 20 years (Table 8.1). The bottom panel of Figure 8.1 shows the corresponding trend for the ECMWF probabilistic forecasts: more precisely, it shows the forecast time when the continuous ranked probability skill score of probabilistic forecasts of the 850 hPa temperature over the Northern Hemisphere falls below 0.25. Also this figure shows an upward trend: for example, we can see that the 0.25 value was crossed in 2002 at forecast day 6.6 and in 2022 at forecast day 9.3, thus indicating a gain in predictability of 2.7 days over 20 years (Table 8.1). These two metrics are part of the headline scores used by ECMWF to track its performance (similar conclusions could be drawn by considering another threshold and/or another field to represent the extra-tropics, large-scale flow).

Figure 8.2 shows the trends in the forecast skill of a more difficult variable to predict, the 24-hour accumulated precipitation over the extra-tropics (the area north of 30°N and south of 30°S). This variable is more difficult to predict because it is more affected by small-scale features than the 500 hPa geopotential height or the 850 hPa temperature. The top panel of Figure 8.2 shows the forecast time when the Stable Equitable Error in Probability Space Score (SEEPS) of ECMWF single high-resolution forecasts of the 24-hour accumulated precipitation falls below a threshold: more precisely, when (1-SEEPS) falls below the 0.45 value. The bottom panel of Figure

Table 8.1 Forecast lead time when the forecast skill dropped below the thresholds highlighted in Figures 8.1 and 8.2, for ECMWF single and ensemble forecasts at the beginning and at the end of the time periods shown in the figures

	Large-scale flow			Small-scale features (Rain)		
	2002	2022	Predictability gain	2003	2021	Predictability gain
HRES	5.5	6.9	1.4d / 20y (0.70 d/de)	2.8	4.8	2.0d / 18y (1.11 d/de)
ENS	6.6	9.3	2.7d / 20y (1.35 d/de)	3.8	7.1	3.3d / 18y (1.83 d/de)

For the large-scale flow, the lead times in 2002 and 2022 have been considered, while for total precipitation, the 2003 and 2021 values have been considered. The predictability gains have been defined as the differences between the lead times, and are expressed in days per decades (d/de).

8.2 shows the forecast time when the continuous ranked probability skill score of ECMWF ensemble probabilistic forecasts of total precipitation falls below 0.10. The lines show the 12-month running average. Forecasts have been verified against station observations. Table 8.1 lists the forecast times when the two thresholds are crossed in 2003 and 2021, and the gain in predictability over this 18-year period. These two metrics are also part of the headline scores used by ECMWF to trace and report its forecast performance.

Although these are only four examples, they provide a good overview of how forecasts have been improving in the past 40 years at ECMWF, the leader in medium-range weather prediction. Operational weather prediction centers use a wide range of scores to assess more specific forecasts: ECMWF, for example, publishes more than 70 scores on their website devoted to the quality of their forecasts, and they mainly focus on eight of them (the headline scores) to monitor routinely forecast quality.

Table 8.1 lists the gains in predictability deduced from the four metrics shown in Figures 8.1 and 8.2. The table shows that, in the medium range (i.e., considering forecasts valid for days 3–15), single forecasts have improved by about 1 day per decade, and probabilistic forecasts by about 1.5 days per decade.

Hereafter are three further indications of forecast quality improvements in the prediction of extreme weather events in the medium-range and large-scale events in the extended (monthly) and long (seasonal) time ranges:

- *Tropical cyclone track prediction*—The single high-resolution forecast position error of a 3-day forecast of the most advanced operational weather prediction centers was about 500 km in 1992, and it is now between 200 and 300 km.

- *Monthly prediction of organized convection in the tropics*— The forecast lead time of the ECMWF prediction of organized convection in the tropics (Madden-Julian Oscillation) increased from forecast day 15 in 2002 to forecast day 27 in 2013. This is equivalent to a predictability gain of about 12 days in a decade; a similar gain has been shown by other operational monthly prediction systems.

- *Seasonal forecast prediction of sea surface temperature in the Pacific*—The lead-time at which the ECMWF forecast of the sea surface temperature in the El Niño 3.4 area in the Pacific drops below 0.9 has increased from about 2.75 months to about 4.5 months between 1997 (Seasonal System-1) and 2017 (Seasonal System-5). This is equivalent to a predictability gain of 28 days in a decade; similar gains have been shown by other seasonal prediction systems.

8.2 How did we manage to improve the accuracy and reliability of weather forecasts?

Progress in four key areas has helped us to improve weather forecasts: people, who have advanced the science and understanding of weather phenomena; observations, which have increased in number and accuracy; supercomputers, which have made it possible

to numerically integrate more complex models; and investments,
which have allowed all of the above to be realized.

People, observations, supercomputers, and investments: the combination of advances in these four areas has made it possible to advance weather prediction.

Thanks to more and more accurate observations, we have understood better physical processes, and we can now estimate more accurately the state of the Earth system, including the initial conditions needed to generate weather forecasts. But observations alone would not have allowed us to make progress: initial conditions are more accurate also because better models have allowed us to use them in a better way, and the availability of more computer power has allowed us to develop more accurate data assimilation procedures capable of assimilating more observations. All of these advances would not have been possible without people, who have developed the models and understood the science, and investments.

We have seen earlier that in the atmosphere all spatial and temporal scales matter, and small errors in some local areas could propagate and affect the global forecast quality. Thus, to improve forecast accuracy, we need to improve key relevant processes already included in the model or add processes that are not yet simulated.

Figure 8.3 illustrates two examples of model improvements. The first one (top panel of Figure 8.3) refers to the role of improvements in air-sea interaction processes on the prediction of tropical cyclones, and larger-scale flows and sea surface temperature anomalies. The diagram shows five processes, characterized by different spatial and temporal scales: air-sea interaction and local weather (e.g., the presence of small islands, and land-sea contrasts) that occurs on local and fast (X_S, T_S) tropical cyclones characterized by relatively larger and slower scales (X,T), and large-scale atmospheric flows and sea surface temperature anomalies that occur on even larger and slower scales (X_L, T_L). Solar radiation is also shown in Figure 8.3 since it provides the energy that drives all processes. Arrows

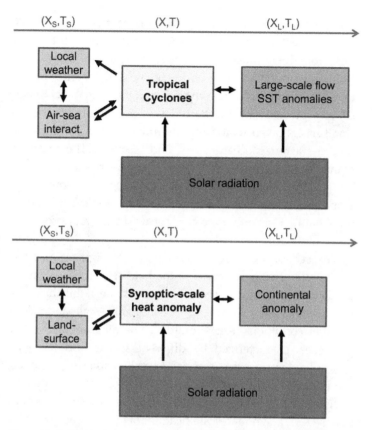

Figure 8.3. Schematic of the role of processes with different spatial and temporal scales [spatially smaller and faster, (X_S, T_S); spatially larger and moderately slow, (X,T); and spatially even larger and much slower, (X_L, T_L)], for the prediction of tropical cyclones (top panel) and of synoptic-scale heat anomalies (bottom panel). See text for more details.

indicate interactions between the different processes: for example, tropical cyclones are affected by air-sea interactions (e.g., exchanges of heat and momentum via ocean waves) and affect the state of the sea; the propagation of a tropical cyclone is affected by the large-scale atmospheric flow and the sea surface temperature, and vice versa, the passage of a tropical cyclone affects both of them. One reason why tropical cyclone predictions have improved substantially in the past 20 years is a more realistic simulation of the air-sea interaction, thanks to improvements of the ocean wave model and to the coupling to a three-dimensional model of the ocean. Observation campaigns have allowed us to diagnose the model performance and address its weaknesses, and the availability of new sources of data (e.g., scatterometer observations of the sea state) has allowed us to properly initialize also the variables that characterize ocean waves and currents. The improvement in the prediction of tropical cyclones has also induced improvements of the simulation of the large-scale flow, for example, over Europe, when there are active tropical cyclones in the Atlantic Ocean that are curving westward and propagate northwestward.

The second example (bottom panel of Figure 8.3) refers to the role of improvements in the land-sea processes on the prediction of synoptic-scale heat anomalies; this has also contributed to improvements in continental-scale anomalous weather conditions (e.g., droughts). The diagram shows five processes, characterized by different spatial and temporal scales: land surface processes and local weather (e.g., clouds, the presence of local water pools/rivers, of vertical motions triggered by local orography) characterized by small and fast scales (X_s, T_s), synoptic-scale heat anomalies characterized by relatively larger and slower scales (X, T), and continental-scale phenomena that occur on even larger and slower scales (X_L, T_L). Solar radiation is shown again since it provides the energy that drives all processes. Arrows indicate interactions between the different processes: for example, land surface processes determine evapo-transpiration, which can lead to cloud formations when there is enough humidity in the soil and thus influence the surface temperature. This can affect the synoptic-scale flow: it can lead to sustained blocked conditions, which can cause a further intensification of the land surface warming. This can further extend the area characterized by an atmospheric blocking condition, as it happened, for example, in summer 2003 and in summer 2022 in Europe.

One reason why heat-wave predictions have improved substantially in the past 20 years is a more realistic simulation of the land surface processes, including the role of vegetation, which have improved the simulation of the energy and humidity fluxes between the soil and the atmosphere. Observation campaigns have allowed us to diagnose the model performance and address its weaknesses, and the availability of new sources of data has allowed us to properly initialize also the variables that characterize the land surface scheme. The improvements in the prediction of synoptic-scale heat anomalies have also led to improvements of the large-scale flow, and in particular of the intensity and duration of continental-scale heat waves and drought conditions.

People working at academic institutions, national meteorological centers and research institutes, and international organizations, by sharing their ideas, knowledge, and processes, have allowed science to advance rapidly. The nature of the numerical weather prediction problem, its global reach, makes it impossible for individual institutions or nations to advance without interacting with each other.

8.3 Can we visualize in a single diagram our prediction capabilities?

Yes, we can use the forecast skill horizon diagram to visualize how far ahead we can predict weather phenomena today, and the dependency of forecast skill on the spatial and temporal scale of the weather phenomena we aim to predict.

Since forecast skill depends on the spatial and temporal scale of the phenomena we are trying to predict, an effective way to visualize how far ahead we can predict weather events today is to use a two-dimensional diagram that has the spatial and temporal scale of the events on the x-axes, and the forecast lead time up to which a skillful forecast can be issued on the y-axes. This diagram is shown in the top-left panel of Figure 8.4. Note that we have used a logarithmic scale on both axes. On the x-axes, spatially small and fast scale events (e.g., heavy precipitation events linked to localized convection) are represented close to the origin, and the broader and slower phenomena (e.g., continental-scale heat waves or large-scale flow features such as extra-tropics and high-pressure systems) are represented at the far end of the axes.

We can start populating this diagram by marking, in gray, the scales that are not resolved by our forecast model: for example, in the top-right panel of Figure 8.4 that illustrates the situation in the 1980s, the gray area extends up to about 200

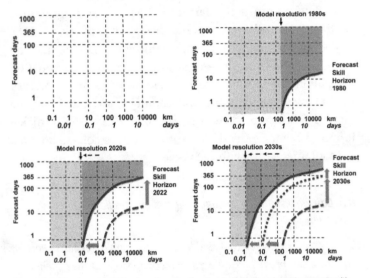

Figure 8.4. The forecast skill horizon diagram visualizes how far ahead weather events with different spatial and temporal characteristics can be predicted. The top-left panel shows the two axes of the diagram: the x-axes indicate the spatial and temporal scale of the phenomena, and the y-axes shows the forecast time (in days) up to which a skillful prediction can be issued. The top-right panel shows the forecast skill horizon at the beginning of numerical weather prediction, in the 1980s, when forecast models had a resolution of about 200 km. The gray area highlights the scales that are not resolved by the model, the dark gray area shows the unpredictable region, and the forecast skill horizon is shown by the solid black thick line. The bottom-left panel shows the forecast skill horizon at the time of writing (2022), when global model resolution has reached about 10 km and the move from atmosphere-only to Earth system models have made skillful monthly and seasonal forecasts possible. The bottom-right panel shows how the forecast skill is expected to expand in the forthcoming decade.

km, which was the grid spacing of the models used in the 1980s. We can then mark in dark gray the unpredictable region, which limits the forecast time up to which different scales can be predicted. In the 1980s, numerical weather prediction was at its infancy, and the work of Edward Lorenz suggested that because of the chaotic nature of the atmosphere and the butterfly effect, the forecast skill limit was about 2 weeks. In one of his works, Lorenz stated that weather forecasts beyond 2 weeks seemed impossible. The thick black line that separates the white area from the gray and the dark gray areas represents the forecast skill horizon estimated at that time. It extended from a few days for phenomena with a spatial scale of few hundred kilometers and a characteristic time scale of a few days, to 10–12 days, for large-scale patterns with a continental scale and a characteristic time of 5–10 days, such as extra-tropical blocking events.

The other two panels of Figure 8.4 show the diagrams at the time of writing (2022) and how it could evolve in the next decade. In 2022, the highest resolution global forecast models have a grid spacing of about 10 km, and some of them generate skillful forecasts up to about 1 year. The leading operational weather prediction centers use Earth system models, capable of simulating realistically also slowly evolving, large-scale phenomena linked, for example, to sea surface temperature anomalies (e.g., linked to organized tropical convection, or El Niño or La Niña events) or to soil anomalies (e.g., linked to reduced soil moisture). Compared to the 1980s, resolution increases and model improvements have led to a "retreat" of both the gray and the dark gray areas, and the expansion of the white area. Increased resolution has made it possible to predict smaller spatial scales. Proper initialization and model improvements have also led to an extension of the forecast skill of the synoptic scales that were already simulated in the 1980s. The use of Earth system models and the adoption of ensemble-based probabilistic approaches have led to an extension of the forecast skill horizon for the large-scale events past 2 weeks, past what was considered the predictability limit in the 1970s.

The 2022 forecast skill horizon curve has been drawn by considering forecast quality reports by the operational weather centers, and studies published in the scientific literature, including the following:

- Small-scale precipitation events (with a spatial scale of 10–50 km) can be predicted up to 2–3 days ahead.
- Extra-tropical wind storms (with a spatial scale of 20–100 km) can be predicted up to 3–5 days ahead.
- Tropical cyclones (with a spatial scale of 100–1,000 km) can be predicted up to 7–10 days ahead.
- Large-scale synoptic features (with a spatial scale of 500–2,000 km) can be predicted up to 2–3 weeks ahead.
- Organized tropical convection in the tropics (with a spatial scale of 1,000–2,000 km) can be predicted up to about 4 weeks ahead.
- Large-scale phenomena linked to sea surface temperature anomalies in the tropical Pacific (with a spatial scale of 5,000–10,000 km), e.g., associated with the El Niño/La Niña events, can be predicted up to 1 year ahead.

The bottom-right panel of Figure 8.4 shows how the forecast skill horizon could be further extended in the next decade if we continue with the improvement trends of the last two decades, which could lead to the grid spacing used in global models to decrease from about 10 to about 1 kilometers, and to more skillful monthly and seasonal forecasts.

8.4 Why does the forecast skill depend on the phenomena we are trying to predict?

There are at least three key reasons why the forecast skill depends on the phenomena that we are trying to predict: errors that affect the small and fast scales grow faster than errors that affect the large and slow scales; there are more uncertainties in the initial conditions of the small and fast scales; and model uncertainties are also larger in the simulation of the small and fast scale processes,.

Figure 8.5 shows a schematic of the spatial and temporal characteristics of some weather phenomena, starting from turbulence (e.g., vortices in the atmosphere caused by the interaction of the atmospheric flow with orographic features), characterized by the smallest spatial and fastest scales, to single convective clouds, organized convection, tropical waves, and planetary waves, characterized by the broadest and slowest spatial scales. Each ellipsoid indicates the range of the spatial and temporal characteristics of the phenomena: for example, single convective clouds have spatial scales between a few hundred meters to a few kilometers, and temporal characteristic times of between a few seconds to a few hours. Note that the ellipsoids align along the diagonal of this graph, meaning

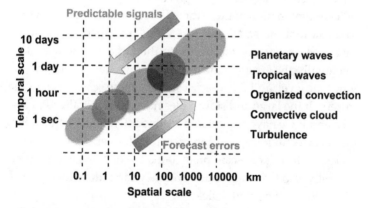

Figure 8.5. Schematic of the characteristic spatial scale (x-axes) and temporal scale (y-axes) of five weather phenomena: turbulence, convection, organized convection, tropical waves, and planetary waves. The bottom arrow indicates the predominant route of propagation of forecast errors that start affecting first the small and fast scales, and then the larger and slower scales. The top arrow indicates the fact that predictability is longer for the large and slow scales, and that predictable signals in these scales can lead to longer predictability also for smaller and faster ones.

that phenomena characterized by relatively short (large) spatial scales are also characterized by relatively fast (slow) characteristic times.

One of the key reasons why the forecast skill depends on the phenomena that we are trying to predict is that the propagation speed of the errors that affect the small and fast scales is fast, and the error reaches a saturation level earlier than the errors that affect the large and slow scales. This has been proven by measurements and shown in numerical experiments.

Another reason why the forecast skill depends on the phenomena that we are trying to predict is that there are larger uncertainties in the initial conditions of the small and fast scales: this is partly because observation errors affect the smaller scales more than the larger scales.

A third reason is that model errors affect the small and fast scales more: for example, the scales linked to turbulence, convection, or the interaction between incoming solar radiation and clouds' layers are more difficult to simulate than larger scales, such as large-scale troughs or high-pressure systems.

The combination of these three factors causes forecast errors to affect first the small and fast scales, and then, as the forecast time progresses, to affect also the large and slow scales. Eventually, the errors in the large and slow scales reach the asymptotic level, and the forecast skill for these scales also drops to zero. If the large and slow scales have been initialized properly, and if we have a good forecast model, then we can evolve them correctly for a long forecast time, longer than the small scales. Skillful large-scale predictions are necessary to predict skillfully also the small scales, and they can induce longer predictability in the small scales.

This contrast between the upscale error propagation and the downscale propagation of predictable signals is represented in Figure 8.5 by the arrows. The bottom arrow is darker in color for the fastest and smallest scales, to indicate that errors grow faster in the small scales and then propagate toward the larger and slower scales. The top arrow is darker in color for the larger and slower scales, to indicate that predictability is longer for these scales.

The forecast skill horizon shown in Figure 8.4 is determined by the envelope of all the forecast times when the errors overcome the predictable signals of all the phenomena (characterized by the different spatial and temporal scales).

8.5 Are extreme events more difficult to predict than the "normal" weather?

Extreme events are more difficult to predict, in terms of their positioning and their intensity.

With the term "extreme events" we refer to events that are very intense and can cause damage and the loss of life: events characterized by extreme values in terms of temperature (heat or cold waves), precipitation (leading to flooding or to drought conditions), or wind (tropical storms, winter extratropical storms). With respect to climatology (i.e., the statistics of weather events of a season), the extreme events populate the tails of the distribution (i.e., they are rare). Because of their intensity, models can have difficulties in predicting precisely how extreme temperature, precipitation, or wind can be, especially when the intensity is due to small and fast scale phenomena.

Let us consider a few extreme weather phenomena and briefly summarize how far ahead we can predict them:

- *Tropical storms*—They have a rather large spatial scale, say between 50 and 500 km. The existing global models are able to predict their path rather accurately, although in some cases they still have difficulties in predicting how deep the pressure at the center of the storm can be, and as a consequence they underestimate the strength of the wind (that causes most of the damages) and the intensity of the associated precipitation. Thus, while the general structure of the tropical storm can be captured quite well many days in advance, the detailed, local precipitation and wind intensity is still affected by errors. Today ensemble strike probabilistic forecasts are accurate and reliable up to 7 to 10 days ahead.

- *Tornadoes*—They have a rather small spatial scale, say from a few hundred meters to a few kilometers, and also a rather fast spatial scale (a few hours). Today, global ensembles with resolutions of order 10 km can only predict the probability that intense wind storms can affect a certain region, but they cannot predict the formation and position of the vortices that characterize tornados. Operational limited-area models with a grid spacing of 1–2 km can simulate them.

- *Heat and cold waves*—They have a large spatial scale and affect large regions (a few hundred kilometers) for several days, even weeks. Today, ensemble probabilistic forecasts can predict them 2–4 weeks in advance, although capturing exactly the intensity of the local temperature associated with heat/cold waves is still challenging. Errors in this case can be linked to the fact that the local orography and soil characteristics (including the presence and characteristics of cities) are not well captured by the models. This latter weakness can lead to errors in the local exchange of humidity and heat between the soil and the atmosphere, affecting the quality of the predicted intensity. Errors can also be linked to poor predictions of the local cloud cover, which can also lead to a wrong estimation of energy fluxes throughout the atmosphere.

- *Flash floods*—They have a very small spatial scale (say, a few kilometers) and are caused by extremely intense, local precipitation, often due to the interaction of the atmospheric flow with the local orography. They are also very fast events: they develop in a few hours and can lead to a hundred millimeters of rain (say, the equivalent of a month of average rain) falling in one place very abruptly. Global ensembles can give reliable probabilities that extreme precipitation events could occur in a certain region, but they still have difficulties in predicting precisely their localization and intensity. Limited-area ensembles with a resolution of 1–2 km can provide more accurate and reliable forecasts than global ensembles.

- *Floods*—They have a large spatial scale (say, from tens to hundreds of kilometers) and are caused by above-normal precipitation affecting a large area for many days, or by extreme precipitation events that affect an area that had already seen above-normal precipitation, and thus have a saturated soil. Events that are characterized by a strong large-scale forcing (e.g., due to the presence of a quasi-stationary trough that leads to a continuous flow of moist air in a region already affected by strong rain) can be predicted 2–3 weeks ahead, precisely because they are linked to slow-moving, large-scale atmospheric features. But in general, correctly predicting the intensity of these events beyond 5–7 days is still very challenging.

- *Droughts*—They have a large spatial scale (say, from tens to hundreds of kilometers) and are caused by an absence of precipitation affecting a large area for many days. They are often combined with periods of very high temperatures: the combination of a lack of rain and high temperature leads to the gradual evaporation of all the deep-soil moisture, eventually leading to drought conditions. Today, they can be predicted 3–4 weeks in advance, although, as it is the case for heat waves, the local intensity of drought conditions could be misrepresented by the models, because of the lack of a precise knowledge of the soil characteristics that determine the heat and moisture fluxes.

- *Extra-tropical wind storms*—They are the equivalent of tropical storms for the extra-tropical latitudes, but compared to them, they have a smaller scale (say, between 25 and 250 km) and are characterized by a slightly shorter characteristic time (they last a few days, while tropical storms can last up to 5–10 days). Today, they can be predicted 3–5 days ahead, although predicting their intensity and local characteristics is usually rather challenging, again because models have difficulties in simulating the local, small–scale features of the wind and the rain associated with the storms.

8.6 What is the minimum spatial scale that a model can simulate realistically?

The minimum spatial scale that a model with a grid spacing Δx can predict is about 4 times Δx ; thus, a model with a grid spacing of 10 km can simulate realistically phenomena with a spatial scale coarser than about 40 km.

Numerical models solve the primitive equations (used to generate weather predictions) numerically, approximating the derivatives with final differences. Approximating the derivatives with finite differences introduces errors that can grow in time: how fast these errors grow depends on the numerical methods. The ECMWF model, for example, uses a combination of spectral and grid-point methods, whereby part of the equations are solved using spectral techniques, and part are solved using finite differences. Other models do not use spectral methods and solve all the components of the model equation using finite difference methods only.

The choice and the design of the numerical methods play a very important role in numerical weather prediction. One of the constraints that all schemes have to take into account is that the time step used to approximate the time derivatives must be less than the time that the smallest wave simulated by the model takes to move from one grid point to the next: this is called the Courant-Friedrichs-Levy (CFL) stability condition. The CFL stability criteria imply that, when resolution is increased and grid points get closer, the integration time step is shortened. The way grid points are organized on the three-dimensional mesh also influences the minimum scale that a model can simulate realistically: some grids have the points organized at a regular distance on both the latitudinal and longitudinal direction, while others position the points following other geometrical shapes, for example, in triangles or octahedrals.

The choice of the numerical method and the way the grid points are organized on the three-dimensional grid have an impact on the capability of a model to simulate realistically phenomena with different temporal and spatial scales, where with the term "simulate realistically" we mean that, on average, the total energy of all the waves simulated by the model is similar to the total energy of all the waves in the real atmosphere.

Figure 8.6 shows a measure of the energy of the waves simulated by two versions of the 2016 ECMWF model (dashed gray lines) and of the waves in the real atmosphere (back solid line). The measure of the energy is the product between the kinetic energy of the waves multiplied by the total wave number of the wave to the 5/3 power. The two ECMWF model versions have the same grid spacing (10 km), but they use a different grid: the dark gray model uses a linear grid, and the light gray model

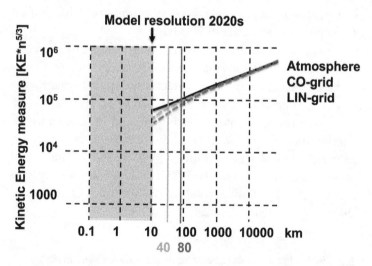

Figure 8.6. Comparison between a measure of the kinetic energy of the real atmosphere (black solid line), against the ones of two model versions with a 10 km horizontal spacing: one with a linear grid (dashed dark gray line) and one with a cubic-octahedral grid (dashed light gray line).

uses a cubic-octahedral grid, which has grid points disposed in octagonal shapes. The cubic-octahedral grid includes about 30% more grid points than the linear grid: since the computational cost of a numerical integration depends on the number of grid points, the model with the cubic-octahedral grid costs more than the model with the linear grid.

Note that both dash curves start diverging from the black curve at a certain threshold, which is at a grid spacing of about 40 km for the cubic-octahedral model (light gray dashed line) and at a grid spacing of about 80 km for the linear-grid model (dark gray dashed line). This means that, although both model versions are able to generate waves with a spatial scale finer than this threshold, they cannot represent these waves with the same energy as in the real atmosphere. In the model simulations, they have a lower energy and a lower intensity, and they propagate slower than in reality. This is why we say that a model with a grid spacing Δx in physical space can only resolve realistically waves with a grid spacing of about $4 * \Delta x$ (the precise multiple depends on the numerical scheme used to numerically integrate the model).

This is the main reason why models have difficulties in simulating the fast and smaller spatial-scale phenomena, even if these phenomena have a characteristic spatial scale that is 2–3 times coarser than the model spatial scale.

By increasing the model resolution, and in parallel by reducing the integration time step (to satisfy the Courant-Friedrichs-Ley stability condition), we can reduce the spatial scale at which the model and the real kinetic energy spectra starts diverging, thus expanding the spectrum of the scales that are simulated realistically.

8.7 What is the minimum spatial scale properly resolved in data assimilation?

The minimum spatial scale properly resolved in data assimilation associated with a model with a grid spacing Δx is a few times Δx : this *is partly because the model simulates realistically only scales up to about 4 * Δx , and partly due to data assimilation approximations.*

Data assimilation adjusts a first-guess state of the atmosphere to be as close as possible to all available observations within a certain time window. The assimilation time window is usually set to span a few hours, say between 3 and 12 hours, and the first guess is usually provided by the most recent short-range forecast.

Figure 8.7 is a schematic of the data assimilation process followed to compute the initial conditions (also called the analysis) of a numerical integration. Suppose that our data assimilation works with a 12-hour assimilation window, and

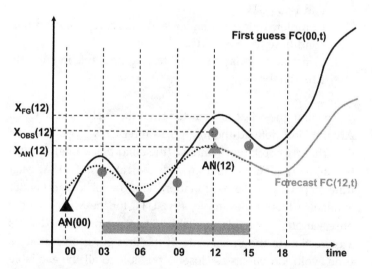

Figure 8.7. Schematic of a data assimilation process with a 12-hour assimilation window, seen on a two-dimensional plane with time on the abscissa and a variable X on the ordinate. The first guess FC(00,t) is a forecast that started from the analysis at 00 UTC, AN(00) (the solid black line). At each time step within the assimilation window (03.00 to 15.00 UTC), observations are collected (dots). The data assimilation procedure defines a correction to the initial state of the first guess such that the new forecast passes as close as possible to all the available observations (the dotted black line). In particular, at 12 UTC, the new forecast that starts from the corrected initial conditions has a value $X_{AN}(12)$ [instead of $X_{FC}(12)$]. $X_{AN}(12)$ is the analysis at 12 UTC: from this analysis, the forecast FC(12,t) is generated. $X_{FG}(12)$, $X_{OBS}(12)$, and AN(12) denote the first guess, the observation, and the analysis value at 12 UTC.

that we produce daily an analysis at 00 and one at 12 UTC (Coordinated Universal Time). Suppose that we issued our last forecast at 00 UTC, it is now 16.00 UTC, and by 19.00 UTC we want to issue a new weather forecast (that starts at 12 UTC). Thus, we need to compute the initial conditions for our 12.00 UTC forecast, and to do this, we run our data assimilation procedure over a time window from 03.00 and 15.00.

These are the key steps that we need to complete to generate the initial conditions and issue a 15-day forecast:

a. Have access to a first guess, in this case a +15-hour forecast started at 00 UTC.
b. Collect all available observations taken between 03.00 and 15.00 UTC.
c. Complete data assimilation and thus compute the initial conditions at 12.00 UTC.
d. Generate a 15-day forecast starting from this initial state (the "analysis").

The first guess $FC(00,t)$ started from the analysis at 00 UTC, $AN(00)$, and defined the first-guess forecast $X_{FG}(t)$ that spans the whole assimilation window (the solid black line): in particular, at 12 UTC the first guess has a value $X_{FG}(12)$. At each time step within the assimilation window, observations are available (dots): for example, at 12 UTC for the X variable the observation $X_{OBS}(12)$ is available. Data assimilation defines a correction to the initial state of the first guess such that its time evolution passes as close as possible to all the available observations within the assimilation window. Once this correction has been computed, a new forecast trajectory (the dotted black line) is generated: this trajectory passes closer to all the available observations than the first guess. In particular, at 12 UTC, the new forecast trajectory has a value $X_{AN}(12)$, instead of $X_{FG}(12)$: $X_{AN}(12)$ defines the analysis at 12 UTC. This new state is used as the initial conditions from where the new forecast $FC(12,t)$ is generated.

From a computational point of view, steps b to d cannot take more than a few, say 3, hours to complete, since by 19.00 we have to issue our new forecast: this limits the resolution of the assimilation procedure and the number of observations that can be used. A further complication arises if we want to generate an ensemble of N forecasts instead of only a single forecast, since we need to generate N initial conditions, and not just one.

Step b involves collecting many observations taken within the assimilation window (03–15 UTC in this case): at ECMWF, for example, about 300 million observations are collected within a 12-hour time window. Note that the data assimilation procedure that covers the 03–15 UTC time window started at about 16.00 UTC, to allow 1 hour for the observations collected toward the end of the assimilation window (03-15 UTC) to arrive at ECMWF. As observations arrive, they are quality controlled and thinned to the maximum number that can be used in the data assimilation procedure.

At the time of writing (August 2022), with the supercomputers currently available, it takes about 30 minutes to generate a 10-day forecast with a resolution of about 10 km, and about 30 minutes to generate all weather forecast products. Thus, the analysis valid for 12.00 UTC must be ready by 18.00 if all forecast products have to be issued by 19.00, which means that, if data assimilation starts at 16.00 UTC, we have about 2 hours to complete the data assimilation procedure and generate the analysis.

The fact that we have about 2 hours to complete the data assimilation procedure puts a constraint not only on the number of observations that can be assimilated but also on the assimilation time window. At ECMWF, for example, the 12-hour assimilation time window is also not centered on the analysis time to avoid waiting for too long to receive the observations taken toward the end of the assimilation window. Computer power availability puts a constraint also on the resolution that can be used in data assimilation: at ECMWF, for example, the resolution at which the optimum correction is computed is about 3 times coarser than the resolution of the forecast model.

The combination of these constraints makes the minimum scale properly resolved in data assimilation that uses a model with a grid spacing Δx to be at most 4 times Δx, but it can be even a bigger multiple of Δx.

8.8 How can we further extend predictability?

We can further extend predictability by reducing the initial condition and model uncertainties. We can further refine the temporal and spatial scale of the phenomena we predict by increasing model resolution and making sure to have enough good-quality observations to initialize the smallest scales. Moreover, we can further extend the forecast skill horizon of the phenomena we are already predicting by improving the simulation of existing processes, by including relevant processes that are not yet simulated, and by improving the initialization schemes.

It is via a combination of more realistic models, more and better observations, and reduced initial uncertainties that we can further expand predictability. Consider the bottom-right panels of Figure 8.4: the further expansion of the forecast skill horizon has been drawn assuming that advances were made on all these areas. Increasing only the model resolution without acting on the model could expand the lower part of the forecast skill horizon, but only by improving the model could we also extend the predictability of the larger and slower scales.

When resolution is increased, the model becomes able to explicitly simulate scales that it did not include before. For example, an increase in vertical resolution could have a substantial impact on the way the model simulates clouds, since the model becomes capable of simulating thinner cloud layers that were not simulated before the upgrade. The cloud scheme can then be adjusted to be able to exploit the resolution increase: since clouds affect how energy propagates throughout the atmosphere and reaches the Earth surface, failing to adjust it could lead to biases, which could affect especially the long forecast range (since it might take days before small biases linked to cloud cover errors can have a substantial impact on forecast quality). Thus, paradoxically, without a proper model adjustment, an increase in vertical resolution could lead to deterioration in the long forecast range.

Here is a second interesting example, linked to the early 2010s, when ECMWF improved its convection scheme, which led to substantially better and more realistic precipitation statistics in the tropics and a bias reduction. This had a major positive impact on the model capability to simulate organized convection, a phenomenon that is key to be able to predict correctly the Madden-Julian Oscillation (MJO). Since the MJO affects the extra-tropical large-scale circulation over Europe, improvements in the prediction of the MJO induced an extension of the predictability of large-scale phenomena over Europe. Thus, by improving the simulation of the small and fast scales linked to convection, ECMWF managed to improve also the predictability of slower, large-scale fields in the extended forecast range.

8.9 Key points discussed in Chapter 8 "The forecast skill horizon"

These are the key points discussed in this chapter:

- Weather forecasts are more accurate and reliable than in the past: during the past 40 years of numerical weather prediction, single forecasts improved by about 1 day per decade, and ensemble-based, probabilistic forecasts by about 1.5 days per decade.
- Monthly and seasonal predictions that were thought impossible in the 1960s and 1970s are now feasible for the slow, large spatial-scale phenomena.
- The forecast skill horizon diagram can be used to visualize how far ahead we can predict phenomena with

different temporal and spatial scales, and how model improvements and resolution increases have led to extended predictability.

- The forecast skill horizon is phenomena dependent: on average, fast and small spatial-scale events are more difficult to predict than slow and large spatial-scale phenomena.
- Models with a grid spacing Δx can predict realistically scales only down to about 4 times Δx.
- Data assimilation schemes associated with a model with a grid spacing Δx can also properly resolve only scales with at most about 4 times Δx, but they can be even a bigger multiple of Δx.
- Predictability can be further expanded by reducing initial condition and model uncertainties.

9

CLIMATE CHANGE AND NUMERICAL WEATHER PREDICTION

In this chapter we briefly introduce climate change and discuss links between numerical weather prediction and our understanding, monitoring, and prediction of climate change. More specifically, we will be addressing the following questions:

1. Why should we talk about climate change in this book?
2. What is the greenhouse effect?
3. What is the state of the climate?
4. How much greenhouse gases do we emit in the atmosphere?
5. Is there a link between greenhouse gases emissions and average global warming?
6. Are we responsible for climate change?
7. What are the key sources of uncertainty affecting climate prediction?
8. What do we mean with "initial value" and "boundary condition" problems?
9. Has climate change impacted weather prediction?
10. Has numerical weather prediction helped in understanding climate change?
11. Which aspects of the future climate can we predict?

9.1 Why should we talk about climate change in this book?

We should talk about climate change because, first of all, this is the most urgent, global challenge that humanity has to address, and secondly because the same physical phenomena that determine weather conditions also define the Earth's climate. Furthermore, the same models that are used to predict the weather are used to understand the climate and predict its future evolution. Progress in numerical weather prediction has favored advances in climate science and vice versa.

Advancing our understanding of weather phenomena and designing models capable of predicting them had a major impact on climate science. Reanalyses covering the past decades, especially coupled climate reanalyses, generated using assimilation methods developed to initialize numerical weather prediction models have allowed us to monitor more accurately the evolution of the Earth's climate. Ensemble methods developed to estimate the uncertainty in weather prediction have been used to estimate the probability of future climate scenarios.

Similarly, there has been extremely positive feedback the other way around, from climate science to weather prediction. For example, the development and use of coupled Earth system models to simulate and predict the climate pointed out the potential positive impact that a shift from atmosphere-land models to Earth system models could have had on weather predictions, and indeed today numerical weather forecasts are generated using Earth system models.

9.2 What is the greenhouse effect?

With the term "greenhouse effect" we mean the impact that greenhouse gases accumulated in the atmosphere have on the atmosphere itself and on the Earth's surface, in particular on their temperature.

The Earth's temperature would be about 30°C colder if there were not a greenhouse effect. The problem we face today is not that there is a greenhouse effect, but that too strong a greenhouse effect caused by continued human emissions of greenhouse gases due to the use of fossil fuels has been warming the planet in an unprecedented way. This warming has caused an increased frequency and intensity of extreme weather events, melting of sea ice, melting of glaciers, and a rise in sea level.

To explain the greenhouse effect, let us first consider the following system: the Earth as a solid body that receives heat in the form of solar radiation and emits heat as a black body. In other words, let us neglect the atmosphere. If we apply the first law of thermodynamics, which says that energy in a closed system must be conserved, to this simple system, then at equilibrium, there must be a balance between the absorbed and the emitted energy.

The Earth, as any physical "body" that has a temperature above absolute zero (−273.15°C), emits radiation with certain characteristics (frequency, wavelength, spectrum). The sun, a much warmer body than the Earth, also emits radiation. We can apply the Stefan-Boltzmann law to compute the energy emitted per square meter by a black body with a temperature T.

Assuming that the sun and the Earth behave as a "black body," we can apply the Stefan-Boltzmann law to relate the emitted radiation per square meter to the black body's temperature:

$$E_{BB} = \sigma T^4. \tag{9.1}$$

where $\sigma = 5.67 \cdot 10^{-8} \, Wm^{-2} K^{-4}$.

For the sun, a star whose photosphere has a temperature of about 5,796K [degrees Kelvin, where the x K are equal to $(x - 273.15)$°C, and 0 K is the absolute zero], the amount of energy that each square meter of the photosphere (Figure 9.1) emits is:

$$E_{photo} = 5.67 \cdot 10^{-8} \cdot (5,796)^4 \, Wm^{-2} = 6.4 \cdot 10^7 \, Wm^{-2} \tag{9.2}$$

From E_{photo} we can calculate the energy emitted by the sun—in other words, its luminosity L_o, which is given by the energy

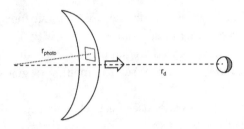

Figure 9.1. Schematic of the calculation of the sun radiation that passes through a square meter at the photosphere, which has a radius r_{photo}, and at the average distance r_d of the Earth from the sun.

emitted per square meter E_{photo} multiplied by the surface the photosphere. Given that the photosphere radius r_{photo} is about 696,000 km, we have:

$$L_0 = 4\pi r_{photo}^2 E_{photo} = 4 \cdot 3.14 \cdot \left(6.96 \cdot 10^8\right)^2 \cdot 6.4 \cdot 10^7 = 3.9 \cdot 10^{26}\,W$$

(9.3)

We can apply equations (9.2) and (9.3) to compute the amount of energy that crosses a square meter at the average distance of the Earth from the sun, which is about 150 million km, by using $r_d = 1.496 \cdot 10^{11}\,m$ instead of r_{photo} in equation (9.3):

$$E_d = \frac{L_0}{4\pi r_d^2} = \frac{3.9 \cdot 10^{26}}{4 \cdot .14 \cdot \left(1.496 \cdot 10^{11}\right)^2} = 1,385\,Wm^{-2}$$

(9.4)

The Earth's surface that receives the radiation from the sun has an area equal to a circle with the Earth's radius, $A = \pi \cdot r_e^2$ (Figure 9.2). If we multiply this area by the amount of radiation that hits a square meter at the average distance of the Earth from the sun, E_d, we can compute the amount of radiation absorbed by the Earth:

$$absorbed\ solar\ radiation = \pi r_e^2 E_d$$

(9.5)

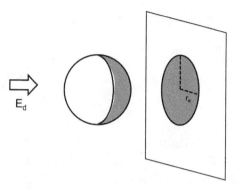

Figure 9.2. Schematic of the calculation of the sun radiation absorbed by the Earth.

To be more precise, we should also consider the fact that part of the solar radiation is reflected by the Earth's surface. This reflection depends on the surface characteristics: for example, the reflection is stronger over surfaces covered by snow or sea ice, and smaller over open sea. On average, by considering the whole Earth's surface characteristics, we estimate that about 30% of all incoming solar radiation is reflected back into space. We can take this into account in equation (9.5) by multiplying the incoming solar radiation by a factor $(1-a_e)$, where $a_e = 0.3$ is the Earth's albedo.

We can now compute the temperature of the Earth, T_e, assuming that it behaves as a black body with a temperature T_e, and assuming that the amount of energy emitted equals the amount of absorbed solar energy, by solving the following equation:

$$absorbed\ solar\ radiation = emitted\ black\ body\ radiation \qquad (9.6)$$

where the emitted radiation is equal to the black-body radiation emitted per square meter times the surface the Earth:

$$emitted\ black\ body\ radiation = \sigma T_e^4 \cdot 4\pi r_e^2 \qquad (9.7)$$

Thus, we can compute T_e by solving:

$$(1-a_e)\cdot \pi r_e^2 E_d = \sigma T_e^4 \cdot 4\pi r_e^2. \qquad (9.8a)$$

$$T_e = \sqrt[4]{\frac{(1-a_e)\cdot E_d}{4\sigma}} = \sqrt[4]{\frac{0.7\cdot 1,385}{4\cdot 5.67\cdot 10^{-8}}} = 255.7\,K. \qquad (9.8b)$$

Equation (9.8b) indicates that 255.7K should be the temperature of the Earth if it behaved as a black body with a 0.3 albedo, and there was no atmosphere around it.

Let us now modify this simple system by adding a uniform atmosphere around the Earth, characterized by a temperature T_A, and that allows solar radiation to cross it without being absorbed, but that absorbs the radiation emitted by the Earth (Figure 9.3). In other words, that includes an atmosphere characterized by a greenhouse effect. This simple atmospheric layer also acts as a black body, and it emits (black body) radiation both toward the free space and toward the Earth surface's that depends on its temperature T_A.

We can write two energy balance equations, one for the atmosphere and one for the Earth's surface, by imposing a balance between the absorbed and the emitted radiation:

$$atmosphere: \; 2\sigma T_A^4 = \sigma T_e^4. \qquad (9.9a)$$

$$surface: \; \frac{1}{4}(1-a_e)E_d + \sigma T_A^4 = \sigma T_e^4. \qquad (9.9b)$$

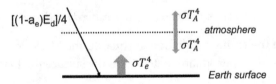

Figure 9.3. The greenhouse effect, explained using a simple model with a uniform, one-layer atmosphere that allows the incoming solar radiation to reach the Earth's surface, and absorbs the outgoing long-wavelength radiation emitted by the Earth's surface.

If we solve these two equations, we find that:

$$T_A = \sqrt[4]{\frac{(1-a_e) \cdot E_d}{4\sigma}} = \sqrt[4]{\frac{0.7 \cdot 1,385}{4 \cdot 5.67 \cdot 10^{-8}}} = 255.7\,K. \qquad (9.10a)$$

$$T_e = \sqrt[4]{2} \cdot T_A = 304.1\,K. \qquad (9.10b)$$

Equation (9.10b) says that, due to the presence of a uniform atmosphere that absorbs the long-wave radiation emitted by Earth's surface and reemits radiation as a black body, the Earth has an average surface temperature of 304K, that is, about 31°C. This temperature is about 50°C higher than the temperature the Earth would have without a greenhouse effect.

Note that 31°C is still not correct, since it is higher than the observed average surface temperature, which is about 15°C. This is because we have assumed that the atmosphere is a uniform, single layer; in other words, because we have oversimplified the Earth and did not consider the complexity of the atmosphere. Despite this weakness, this simple model highlights the important role that an atmosphere that includes greenhouse gases has on the Earth's surface temperature. To make the model more realistic, we could assume that the atmosphere is compounded by a few different layers instead of only one, with each layer characterized by a different temperature, and find a solution for the Earth surface's temperature that is closer to the observed one.

The atmosphere allows most short-wave solar radiation to reach the Earth's surface, while the amount of long-wave radiation emitted by the Earth that it absorbs depends on the concentration of the greenhouse gases that absorb it. The principal greenhouse gases are water vapor, carbon dioxide, methane, nitrous oxide, and fluorinated gases. Table 9.1 lists the main components of the atmosphere: note that the concentrations of the greenhouse gases are very small, ranging from about 0.3% (in mass fraction with respect to dry air) for water vapor to much smaller values for carbon dioxide and methane.

Table 9.1 Atmospheric concentration of the principal greenhouse gases: water vapor, carbon dioxide, and methane

	Total mass (in 10^{21} g)	Mass fraction (with respect to dry air)
Total atmosphere	5.136000	
Dry air	5.119000	
Nitrogen (N_2)	3.870000	75.6007%
Oxygen (O_2)	1.185000	23.1491%
Argon (Ar)	0.065900	1.2874%
Water vapor (H_2O)	0.017000	0.3321%
Carbon dioxide (CO_2)	0.002760	0.0539%
Methane (CH_4)	0.000005	0.0001%

There are two key differences between water vapor and the other greenhouse gases. First, for any air mass, the water vapor concentration strongly depends on the pressure and the temperature of the air. If a mass of air containing water vapor is cooled and/or compressed, water vapor would condense and precipitate; water continues to cycle throughout the Earth system, and as a consequence, the total concentration of water vapor in the whole atmosphere changes very little. Because of this continuous cycle, the time that a molecule of water (H_2O) spends in the atmosphere is about 10 days (compared, e.g., to many decades for carbon dioxide).

Secondly, in the last 100 years humans have increased the concentration of carbon dioxide (CO_2) and methane (CH_4) in the atmosphere. Note that the time that a molecule of CO_2 spends in the atmosphere is between 300 and 1,000 years (it is difficult to give a narrow range, since the residence time estimate depends on assumptions made on the physical processes that dominate the CO_2 removal from the atmosphere and storage), and the time that a molecule of CH_4 spends in the atmosphere is about 10 years. This means that CO_2 and CH_4 continue to exercise their effect long after they have been injected in the atmosphere.

9.3 What is the state of the climate?

The concentration of human-made greenhouse gases, linked to the use of fossil fuels, continues to increase, global warming intensifies, and sea-ice melting and sea-level rise accelerate. In 2022, the CO_2 concentration in the atmosphere reached 418 parts per million (ppm), and the global average temperature was about 1.18°C degrees higher than during the preindustrial time. If we continue with the current level of emissions, global warming will reach 1.5 °C degrees by about 2035, and it will very likely surpass 2.0°C degrees before the end of the twenty-first century.

In March 2022, the atmospheric CO_2 concentration reached 418 parts per million (ppm). Reconstructions of the CO_2 concentrations of the distant past, based on the analysis of sediments and ice cores, indicate that we need to go back about 2.5 million years to detect CO_2 levels above 400 ppm. They also indicate that in the last 800,000 years and up to 1900 the CO_2 concentration oscillated between about 180 and 300 ppm, and that after 1900 it started to climb very rapidly to the current levels. Observations show that not only CO_2 but also the other main greenhouse gases continue to rise: methane (CH_4) has passed 1,900 parts per billion (ppb) and nitrous oxide (N_2O) has passed 335 ppb. In the distant past, variations used to happen on long time scales, of about 100 ppm over periods of 20,000–50,000 years, but never before the industrial time, apart from during catastrophic events such as major volcanic eruptions or meteorites' impacts, has the Earth seen CO_2 variations of about 100 ppm happening over about 100 years, as we have witnessed in recent times, and especially after the Second World War.

Figure 9.4 shows the annual mean concentrations of CO_2 and CH_4, from data collected at the Mauna Loa Observatory (Hawaii, United States). Note that for CO_2 the increase follows an exponential curve, and that the annual percent increase has been growing nearly continuously. In fact, as shown in Figure 9.5, while in the 1960s and 1970s the annual percent increase in CO_2 was between 0.2% and 0.4%, in the last two

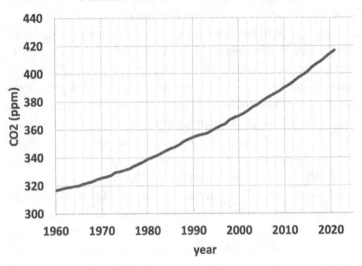

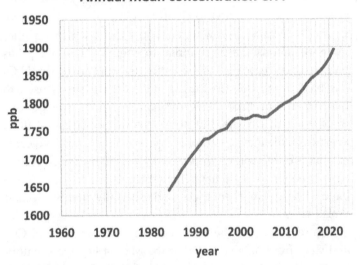

Figure 9.4. Annual mean concentration of carbon dioxide (CO_2, in parts per million, ppm; top panel) and of methane (CH_4, in parts per billion, ppb; bottom panel), measured at the Mauna Loa Observatory. (CO_2 data from Dr. Pieter Tans, NOAA/GML [gml.noaa.gov/ccgg/trends/] and Dr. Ralph Keeling, Scripps Institution of Oceanography [scrippsco2.ucsd.edu/]. CH_4 data from Ed Dlugokencky, NOAA/GML [gml.noaa.gov/ccgg/trends_ch4/])

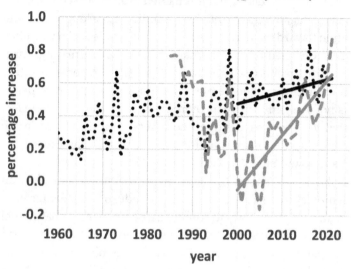

Annual mean concentration increase CO2(black lines) and CH4 (grey lines)

Figure 9.5. Annual percent increase in the concentration of carbon dioxide (CO_2, in parts per million, ppm; black dotted and solid lines) and of methane (CH_4, in parts per billion, ppb; gray dashed and solid lines), computed from the Mauna Loa Observatory data shown in Figure 9.4.

decades it has been above 0.5%. If we consider the last two decades, a linear-fit curve indicates a clear positive trend, with the annual percentage increase going from 0.32% in 2000 to about 0.53% in 2021, with a peak percentage increase of 0.85% in 2016.

CH_4 concentrations have also been increasing from about 1,645 ppb in 1985 (data before 1985 were not available) to 1,879 ppb in 2020 (Figure 9.4). Figure 9.5 shows that for CH_4 the annual growth rate decreased between 1985 and 2000, but since then the CH_4 concentration has been growing at a very fast rate, due mainly to an increased use of methane as a source of energy, and also to the melting of the permafrost that has been releasing increasing amounts of it. The last two decades have seen a clear positive trend, with the growth rate increasing from about 0.06% in 2000 to 0.87% in 2021, an increase of a factor of about 15.

As the concentration of greenhouse gases increased, the atmosphere has been absorbing more long-wave radiation emitted by the Earth, and this has led to an increased warming. A warmer atmosphere has been emitting more long-wave radiation toward the Earth's surface, which has also been warming. This cause-and-effect mechanism between increasing emissions of greenhouse gases and the global average temperature is a direct consequence of the greenhouse effect, which per se is not a negative phenomenon. What is problematic is the fact that we, humans, have caused a very rapid increased in the concentration of greenhouse gases, and this has caused the Earth to warm very quickly.

Figure 9.6 shows the anomaly of the global annual mean temperature with respect to the preindustrial value: for each year, the figure shows the difference between the global annual mean temperature of that year and the global annual mean temperature of the period 1850–1900. The solid line shows the annual values, while the (linear-fit) straight line, which has a slope of ~0.02°C/year, shows the long-term warming trend. Note that superimposed over this linear warming trend of ~0.2°C per decade are natural oscillations of about 0.1°C–0.2°C per year. These natural oscillations are due to internal variability of the atmospheric flow, variations of the heat exchanged between the Earth's oceans and atmosphere, and large volcanic eruptions. For example, large-scale episodes in the tropical Pacific Ocean that cause the ocean temperature to warm (during the El Niño event) or to cool (during La Niña events) produce natural oscillations in temperature.

Figure 9.6 is based on a new dataset produced by the European Union Copernicus Climate Change Service, the ERA-5 reanalysis, constructed by assimilating all available observations of the Earth system in a state-of-the-art model at the European Centre for Medium-Range Weather Forecasts (ECMWF). It covers the satellite era, from 1980 to the present, during which satellite data have allowed scientists to monitor the Earth's temperature very accurately and with many more

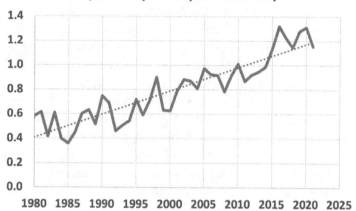

Figure 9.6. Global warming with respect to the preindustrial level. The solid line shows the anomaly of the global annual mean surface temperature with respect to the preindustrial level. The dotted straight line shows the linear fit, which indicates a global warming trend of about 0.2°C every 10 years. (Temperature data from the ERA-5 reanalyses, European Union Copernicus Climate Change Service/ECMWF)

details (e.g., also in regions poorly covered by conventional data) than in the past.

Figure 9.6 shows that in 2016, 2019, and in 2020 the global average temperature was more than 1.2°C warmer than the preindustrial level (defined as the mean temperature between 1850 and 1900), and that the seven years from 2015 to 2021 had been the seven warmest years since 1980.

It must be stressed that, although the globally averaged surface temperature now stands at about 1.18°C above the preindustrial value, some parts of the globe, for example, the polar caps and the Mediterranean region, experience average warming levels much higher than 1.2°C. For example, over Europe the average warming stands at about 2.5°C with respect to the preindustrial level. This means that a further future global average warming of about 1°C degree could translate, for the "climate hot spot regions," into a further warming of at least 2°C degrees.

Climate warming has been causing a sea-level rise and the melting of glaciers and ice caps. Since 2006, an acceleration of almost a factor of 3 of the sea-level rise has been detected, with the average sea level increasing by about 3.6 mm/y, compared to about 1.4 mm/y before 2006. This acceleration is due partly to the warming of the oceans and partly to the melting of glaciers. If we consider the polar caps, observations have been showing that both their extension and thickness have been decreasing: the first months of 2022 have seen the extension of both polar caps below the historical minima detected so far.

The VI Assessment Reports of the Intergovernmental Panel on Climate Change (IPCC) issued in 2021 and 2022 lists the following other key impacts of climate change:

- Climate change is intensifying the water cycle: this brings more intense rainfall and associated flooding, as well as more intense drought in many regions (examples include the floods that affected Central Europe in July 2021, and those in Australia, Bangladesh, and Pakistan in June/July 2022, and the forest fires linked to heat waves and droughts that have been affecting many countries during the past two summers of 2021 and 2022).
- Climate change is affecting rainfall patterns: at high latitudes, precipitation is likely to increase, while it is projected to decrease in the tropics and the Mediterranean region.
- Climate change–induced sea-level rise has accelerated in the past decade, with sea-level rise rates reaching ~3.4 mm/year: extreme events related to sea-level rise that previously occurred once every 100 years could happen every year by the end of this century.
- Climate change–induced warming will amplify permafrost thawing, the loss of seasonal snow cover, melting of glaciers (see, e.g., what has been happening in the Italian Alps, including the disaster of the Marmolada Glacier in July 2022) and ice sheets, and the loss of Arctic sea ice,

which is projected to be ice-free in summer before the end of the century.

- Climate change led to ocean warming, ocean acidification, and reduced oxygen levels that have affected ocean ecosystems.
- For cities, some aspects of climate change may be amplified, including heat waves (since urban areas are usually warmer than their surroundings), flooding from heavy precipitation events, and sea-level rise in coastal cities.

9.4 How much greenhouse gases do we emit in the atmosphere?

In 2019, we have injected into the atmosphere 45 Gton of CO_2-eq (CO_2-equivalent emissions) of greenhouse gases (37 Gton of CO_2 and 8 Gton of methane and other greenhouse gases converted into CO_2-eq by taking into account their long-term impact on the Earth atmosphere). In terms of emissions per capita, between 1990 and 2019 the global annual mean value has increased from 7 to 7.8 tCO_2-eq (tons of CO_2-equivalent emissions), with geographical variations of up to a factor of 10, with some countries characterized by emissions per capita of about and below 2.0 tCO_2-eq, and others by values about and above 20 tCO_2-eq.

The latest report of Working Group III of the IPCC issued in April 2022 states that greenhouse gas emissions continue to grow. It also reminds us that on the one hand greenhouse gases remains in the atmosphere and keep having a warming effect for many decades, and on the other hand that most of the greenhouse gases have been emitted in the last few decades: 17% of all the greenhouse gases injected in the atmosphere between 1850 and today were injected in the atmosphere in the last decade, between 2010 and 2019.

If we consider the emissions of greenhouse gases injected into the atmosphere between 1850 and 2019, the regions that contributed most to the current levels are North America (which emitted 23% of the total amount) and Europe (with 16%), followed by East Asia (12%), Latin America and the Caribbean (11%), and then to the other regions of the world. The last 30 years have seen large variations in the relative contribution of the different regions, as a consequence of economic growth, and the transformation of the economies of many countries that have seen a shift of manufacturing toward the East, and the focusing of some of the Western economies into less-energy-intensive sectors (e.g., financial services). This is confirmed by the fact that between 1990 and 2020, the relative contribution to the global emissions of North America has decreased from 18% to 12%, the European contribution has decreased from 16% to 8%, and the relative contribution of East Asia has increased from 13% to 27%.

9.5 Is there a link between greenhouse gas emissions and average global warming?

Yes, data of the past decades have indicated that there is a quasi-linear relationship between the accumulated emissions of greenhouse gases and the global average warming.

The latest report of Working Group I of the IPCC, published in 2021, for example, reminded us that there is a quasi-linear relationship between the amount of greenhouse gases that are released into the atmosphere and global warming. Figure 9.7, built using global average warming data from the ECMWF ERA-5 reanalyses and accumulated greenhouse gas emission data from the World Bank database, confirms this quasi-linear relationship:

$$T = 0.525 \cdot X_{GHG} + 0.372 \qquad (9.8)$$

where T is the annual average temperature anomaly, and X_{GHG} is the accumulated amount of greenhouse gases emitted in the atmosphere. The coefficient of determination of the linear fit, $R^2 = 0.83$, confirms the robustness of this linear relationship.

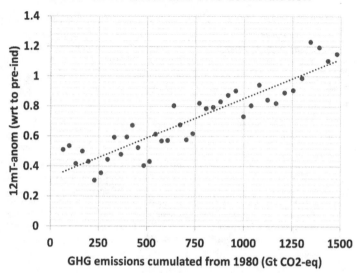

Figure 9.7. Total accumulated greenhouse gases emitted since 1980 (x-axis: data from the World Bank archive) versus global annual average surface temperature anomaly with respect to the preindustrial level (y-axis: data from the European Union Copernicus Climate Change Service/ECMWF).

The fitted straight line shown in Figure 9.7 has a slope of 0.525°C/1,000 Gt, which means that during this period, each additional emission of 1,000 Gt of greenhouse gases in the atmosphere has led to 0.53°C of warming on average.

The variability around the linear-fit straight line reflects the fact that each year's climate is influenced not only by greenhouse gas concentrations but also by internal natural variability, including coupled atmosphere and ocean dynamics (e.g., whether the year was characterized by a strong El Niño or La Niña event, or whether changes in the large-scale circulation caused long-lasting heat waves over large areas of the globe).

We can use the quasi-linear relation of equation (9.7) to estimate how much more greenhouse gases we are allowed to emit in the future if we wish to stay below a given threshold,

Table 9.2 Quasi-linear relationship between accumulated greenhouse gas emissions and global average warming

	GHG accumulated emissions (Gt CO$_2$-eq)	Observed <T>	Linear fit <T> estimate
2000	703	0.6	
2019	1,480	1.2	
2030	1,975		1.3
2040	2,425		1.6
2050	2,875		1.9
2060	3,325		2.2

The first two rows shows the total accumulated emissions (from the World Bank database) and the observed global average warming (from the Copernicus Climate Change Service) in 2000 and 2019. The other rows show the projected total accumulated emissions, calculated assuming that the world continues to emit 45 Gt CO$_2$-eq as it emitted in 2018 and 2019, and the projected global warming computed using the quasi-linear relationship reported in equation (9.7), for 2030–2060.

and how long it will take to surpass some warming limits by taking into account that in 2019 the world emitted about 45 Gt of greenhouse gases. Given that in 2019 the average global warming was about 1.2°C above preindustrial levels, if we want to keep the global average warming below 2°C, we can only emit about 1,500 Gt more greenhouse gases, since emitting 1,500 Gt will induce a further warming of about 0.8°C according to equation (9.7). In 2019, we emitted 45 Gt of greenhouse gases (of which ~30 Gt was CO$_2$ and ~15 Gt was CH$_4$ and other greenhouse gases). If we continue to emit, on average, 45 Gt per year, in 34 years we would reach that amount: thus, in 2055 we would reach the 2°C warming level. If we want to keep global warming below 1.5°C, we have to limit emissions to less than 600 Gt: if we continue to emit as we did in 2019, we will surpass this 600 Gt in about 13 years, at around 2035.

Table 9.2 lists an estimate of the amount of accumulated greenhouse gases emissions and average global warming that we would reach in the forthcoming decades, assuming that the quasi-linear relationship (9.8) continues to hold in the future, and that we continue to emit as in 2019.

9.6 Are we responsible for climate change?

Yes, climate change is attributable mainly to fossil fuel burning and other human activities. The VI Assessment Report of Working Group I of the IPCC published in August 2021 stated very clearly that "It is unequivocal that human influence has warmed the atmosphere, ocean and land. The likely range of total human-caused global surface temperature increase from 1850–1900 to 2010–2019 is 0.8°C to 1.3°C, with a best estimate of 1.07°C."

Here are six facts that support this statement:

1. Analysis of gas bubbles trapped in ice cores extracted from Greenland and Antarctic ice sheet indicates that atmospheric CO_2 started to increase around the time of the Industrial Revolution, and it has roughly tracked the rate of growth of fossil fuel consumption since that time.

2. Atmospheric oxygen (O_2) has been observed to be decreasing at a rate of about 3 ppm per year, consistent with the hypothesis that the CO_2 being added to the atmosphere is a product of combustion.

3. The relative abundance of the radioactive isotope ^{14}C and the stable isotope ^{13}C in atmospheric CO_2 is declining: ^{14}C is virtually absent in fossil fuels and ^{13}C is less abundant in fossil fuels than in atmospheric CO_2 and in carbon dissolved in the oceans; thus the relative decline is due to the increased emission of carbon atoms contained in fossil fuels.

4. Numerical experiments with state-of-the-art Earth system models indicate that the climate evolution of the past century can be reproduced only by considering also human factors, that is, the increase in the concentration of greenhouse gases and of the aerosols. By considering natural factors (i.e., natural variability) alone, we cannot reproduce it.

5. Calculations based on radiative transfer models indicate that human factors (the emission of greenhouse gases and of aerosols) are the principal party responsible for the observed warming.

6. About half of the carbon emitted into the atmosphere has been absorbed by the ocean, and the increasing storage of carbon in the oceans in the form of dissolved CO_2 has been increasing the concentration of H^+ ions, lowering the pH of the sea waters.

9.7 What are the key sources of uncertainty affecting climate predictions?

A climate prediction is an extremely long integration of the Earth-system models that include also time-varying greenhouse gases concentrations. For the very long climate time scale, the sources of uncertainty are not only linked to initial conditions and model approximations but also to future greenhouse gas emissions and aerosols.

Let us assume that we have a very good-quality Earth system model and data assimilation, and that we can generate very good estimates of the state of the Earth system.

On the weather prediction time scale—say, from 1 day to a few years—initial condition uncertainties and model uncertainties play a key role. On this integration time scale, the future emissions of greenhouse gases (and in general variations of the concentration of greenhouse gases in the atmosphere) play a secondary role, while we have indications that aerosols play a role on the monthly-to-seasonal time scale. In the short forecast range (say, a few days), initial condition uncertainties dominate model uncertainties. The two terms become comparable around forecast day 3 to 5, and in the longer forecast range (monthly and seasonal time scale), model uncertainties play a dominant role.

As the forecast range extends beyond 1 year into the climate time scale (say, a few decades), uncertainties linked to greenhouse gas emissions start playing a role that becomes dominant as the forecast range extends past a few years. This is because they can affect the energy balance in the atmosphere and at the Earth's surface and induce variations larger than the variations linked to model uncertainties. Thus, on the climate time scale, there is this extra complexity of estimating how emissions will evolve, as a function of population, technological development, and changes to the Earth system components that absorb part of the human emissions (ocean, vegetation). This is an extra complexity that translates into an extra source of uncertainty. Indeed, there are many research groups that aim to provide reliable estimates on how they could vary in the future, and the implications of these variations on the accumulation of greenhouse gases in the atmosphere.

9.8 What do we mean with "initial value" and "boundary condition" problems?

An initial value problem is one whose solution depends on the initial conditions, and a boundary condition problem is one whose solution depends on boundary conditions.

Weather prediction is considered an initial value problem: predicting the future weather depends strongly on the initial conditions, and having approximate initial conditions reduces the forecast skill. Boundary conditions (e.g., variations in solar radiation or of the concentration of the greenhouse gases) can be kept fixed in time, since their time variations have a secondary, smaller impact. A short-range (up to 2–3 days) forecast is mainly dominated by initial uncertainties, but as the forecast lead time lengthens past a few days, model uncertainties start playing a key role. In the monthly and seasonal time scale, having a good knowledge of the initial state of the slowly evolving components (ocean state, sea ice, deep soil conditions) matter and can lead to skillful predictions (i.e., better than climatology). Thus, generally speaking, we can say that weather prediction is essentially an initial value problem and that in weather prediction, initial and model uncertainties are the dominant sources of forecast errors, and a reduction of their amplitude can lead to forecast improvements.

Climate prediction is instead considered a boundary condition problem, since it is very important to take into account the time variation of boundary conditions. Since in this case we are talking about numerical integrations valid for a few decades, knowing exactly the initial state of the atmosphere has little influence. Having a good knowledge of the initial state of the slow evolving components (e.g., the three-dimensional ocean or the sea ice) matters for the first few years of integrations, but this has little impact on the forecasts valid for the forthcoming decades. For these very long time ranges, instead, knowing the forcing induced by greenhouse gases and aerosols matters more: uncertainties linked to these fields dominate over uncertainties linked to initial and model uncertainties. These fields are considered "boundary fields," since they are imposed (as external forcing fields) using the time integrations. They would be considered internal components if the Earth system model included a simulation of how population would grow and of the impact of human activities on greenhouse gases and aerosols. In the climate forecast range, uncertainties in these boundary conditions dominate, followed by model uncertainties: reducing them could lead to a reduction of forecast uncertainty. Thus, generally speaking, we say that climate prediction is essentially a boundary condition problem.

9.9 Has climate change impacted weather prediction?

Yes, climate change has been having an impact on how certain weather products are generated. Moreover, the study of climate change, and the development of Earth system models capable of simulating its evolution, has helped to advance numerical weather prediction.

With a changing climate, events that were extremely rare a few decades ago have become more frequent and intense. Since some weather products are based on the comparison of the latest forecast of the probability distribution function with the climatological distribution, it is very important to estimate this latter using the data from most recent decades, and not to rely on a climatology computed using data from only distant decades. This can have an impact, for example, when we generate seasonal forecasts of the temperature anomaly of a region, either expressed in terms of the average anomaly, or of probabilities of being in the upper or lower terciles. If the climatology is too old, the forecasts will, for example, always predict warm anomalies, due to climate change, or if forecasts are expressed in terms of return periods, they will provide return periods that are too long. Similar biased forecasts could happen if we aim to predict extreme precipitation events. One way to avoid these problems is to continuously update the computation of the climatological distribution, using, for example, the latest 20 years to define it, rather than using a "frozen," fixed number of years. This is the first impact that climate change has been having on weather prediction.

The second impact has been on modeling. The study of climate change, and the development of Earth system models to monitor, understand, and predict the future climate, has had a major impact on numerical weather prediction. While analyzing the long time integrations of Earth system models generated for climate studies, we found evidence of the importance of coupling a three-dimensional dynamical ocean and a dynamical sea ice, and the inclusion of chemical species and aerosols not only on the long term but also in the earlier part of the climate integrations. This is why we use Earth system models also for weather prediction, in ensemble forecasts valid for the monthly and seasonal scale, and increasingly then also for the medium range. As a result, today the difference between the models used in weather and climate prediction is much smaller than in the past.

9.10 *Has numerical weather prediction helped in understanding climate change?*

Yes, numerical weather prediction, the understanding of weather phenomena, and the development of models capable of simulating and predicting them have led to the development of Earth system models that have then been used also in climate studies. More recently, reanalyses generated using data assimilation methods developed to initialize weather models have helped in monitoring climate change more thoroughly.

There has been a deep cross-fertilization between works done by weather and climate scientists, in the science, the modeling of the Earth system, the establishment of thorough and comprehensive observation networks, and the development of data assimilation methodologies. The fact that today the same Earth system (i.e., coupled atmosphere, land, ocean, and sea ice) models are used in both fields made their testing more thorough. The convergence that we have seen in the past two decades has also provided a more thorough testing ground for these models, since the same model is now tested and has to perform well over both the weather days-to-months time scale and the climate years-to-decades time scale.

Consider, for example, the fact that today a dynamical three-dimensional ocean and sea-ice model developed for climate studies is used also in numerical weather prediction. This has provided the opportunity to assess the quality of the model every single day a forecast was issued: this has helped to spot problems in the ocean and sea-ice schemes, diagnose them, and address them. The use of ocean and sea-ice models in weather prediction has also speeded up the development of coupled atmosphere, land, ocean, and sea-ice data assimilation schemes, capable of providing consistent initial conditions for all the components of Earth system models (atmosphere, ocean, land, and cryosphere). The development and testing of coupled data assimilation schemes have led to a better understanding of the dynamics of air-sea interactions, and this has helped to diagnose and improve the ocean and sea-ice models. Once improved, these components have fed back into the climate models, thus leading to more realistic climate projections.

In the near future, a similar cross-fertilization could lead to the inclusion of the carbon cycle and interactive aerosols' models also in weather models. Since climate change is driven by greenhouse gas concentration, and since aerosols also play a role in determining the warming, these two schemes have been included for years in the models used for climate studies. Evidence has now started to emerge that these two components should be included also in weather models, especially for weather forecasts valid for the seasonal time scale. Furthermore, there is an increasing demand to monitor daily greenhouse gases, especially CO_2 and CH_4, and aerosols, and to predict them on the weather time scale. But before including computationally expensive schemes in weather forecast models that simulate their cycle, we need to understand how to initialize them, and whether we can do this using available observations or we need more ad-hoc and detailed ones. Work is progressing in all these areas.

Another cross-fertilization area has been a more thorough validation and tuning of the climate models. The climate models of today use the resolution that weather models used about 20 years ago (there is a factor of about 10 to 20 between the resolution used in the two). This means that a lot of the testing and tuning of the parameterization required have already been done by the weather scientists. This has clearly helped to speed up the preparation of model versions.

9.11 Which aspects of the future climate can we predict?

We can predict the global trend in temperature and characterize how the global trend translates into regional trends. We can predict how the statistics of weather phenomena (e.g., the frequency and intensity of extremes) will change and how the probability of different possible scenarios will vary accordingly to the greenhouse emission scenarios. We can use limited area models, nested into global models, to translate these probabilities into local weather statistics. All these predictions can only be expressed in probabilistic terms.

If we look at the VI Assessment Reports by the IPCC issued in 2021 and 2022, statements about the future climate are all expressed in probabilistic terms: they indicate possible scenarios and include an estimate of the uncertainty around them. Forecasts cover rather large areas and include few local details, since the uncertainties around them are very large.

By nesting limited area models into global climate projections, we can estimate how global scenarios translate into future regional climate variations, and estimate how the probabilities of events of interest could change (e.g., whether the probability of extreme flood events, heat waves, or droughts could change). Care should be taken in extrapolating local details, since the uncertainty in the global and regional projections are still rather large, and this makes it impossible to extract robust signals on the future detailed local weather.

In the future, we expect to improve the global climate forecasts, especially in the "near-climate range," the range going from year 1 to year 5–10 of the decadal climate projections. In this range, we expect to achieve improvements by working on the initialization of the slowly evolving components (oceans, sea ice, and deep soil) and a more accurate definition of the near-future emission pathways. If computer power availability increased by a factor of 10^3–10^6, we should be able to generate ensemble climate predictions at a much higher resolution than today—say, at 1 km resolution instead of the 50–100 km resolution currently used. The increase in resolution is expected to reduce the overall uncertainty of the climate projections and to provide more reliable predictions of the statistics of extreme weather events (hurricanes, extra-tropical storms, extreme heat waves, and extreme precipitation events). We should also be able to introduce in the Earth system models more realistic and dynamical chemical components, and this could also lead to a more realistic simulation of the two-way interaction between chemical species, clouds, and radiation.

If and when we managed to improve our forecasts of the near-climate range, it will be clearer whether we can aspire to improve also the forecasts for the longer forecast range: today, it is difficult to say when and whether this could be achieved.

9.12 Key points discussed in Chapter 9 "Climate change and numerical weather prediction"

These are the key points discussed in this chapter:

- The greenhouse effect has a major impact on the Earth's climate conditions and regulates the average global temperature.
- The climate keeps changing: the concentration of the greenhouse gases in the atmosphere has reached levels that the Earth has not seen for 2.5 million years; as a consequence, the global average temperature has increased by about 1.2°C compared to the preindustrial level (1950–1900).
- In 2018 and 2019, humans emitted 45 Gt CO_2-eq greenhouse gases into the atmosphere; if we look at the accumulated emissions between 1850 and 2019, North America and Europe are the two top emitters.
- There is a clear link between the increase in the accumulated emissions in the atmosphere and the average global warming.
- We humans are the principal actor responsible for the current situation: this statement is supported by theory, observations, and experiments.
- The main sources of uncertainty in climate predictions are model approximations and emission pathways.
- While weather prediction is considered an initial value problem, whereby initial condition uncertainties have a major impact on forecast quality, climate prediction is considered a boundary condition problem, whereby external forcings (linked to greenhouse gases emissions, population growth, and human impact on land conditions) have a major impact on forecasts.

- There has been a lot of cross-fertilizations between weather and climate science, and the fact that today the same Earth system models are used both in weather prediction and climate studies has led to model improvements and a better understanding of the interactions between different components of the Earth system.
- We can predict the global statistics of the future climate and some regional details.

10

A LOOK INTO THE FUTURE

In this chapter we provide an overview of the main areas of research and development in numerical weather prediction. More specifically, we will be addressing the following questions:

1. What are the focus areas of research in numerical weather prediction?
2. What is an Earth digital twin?
3. Will we be able to continue to improve the quality of weather forecasts?
4. Will we ever be able to issue a "perfect" forecast?
5. In 2050, will we be able to predict the local weather of the next season?
6. Can artificial intelligence lead to improved predictions?
7. What is an "environmental prediction model"?
8. Is weather prediction evolving into environmental prediction?
9. As global models keep increasing resolution, will we still use limited-area models?
10. Would a future operational suite look very different from today's?

10.1 What are the focus areas of research in numerical weather prediction?

Research and development to further extend the weather forecast skill horizon is focusing on the following key areas: a further expansion of the range of forecast variables and products; the adoption of strongly coupled Earth system approaches also in data assimilation; the adoption of seamless ensembles in data assimilation and prediction; and the use of higher resolution.

Results obtained in the past two decades have shown that adding relevant processes and adopting increasingly more realistic Earth system models can improve forecast quality and further extend the forecast skill horizon. As more processes are added, new types of observations can be assimilated, and more variables can be properly initialized. The inclusion of more model processes also expands the range of variables that can be predicted in a realistic way: think, for example, of aerosols or chemical species. As new variables are simulated realistically, new products can be designed and offered to the forecast users.

One area where we expect major advances is coupled data assimilation: today, initial conditions are still mainly computed using uncoupled or weakly coupled methods, where with "weakly coupled" we mean that only part of the information contained in the observations taken in one model component is passed to other components during the assimilation process. By moving toward strongly coupled data assimilation, we could better initialize the models and thus reduce the initial uncertainties: for example, we could use a better way to obtain observations of the lower atmospheric layers to initialize the ocean and sea-ice state, and vice versa.

One of the key questions that we need to address in this area is the definition of the weights given to the variables of the different Earth system components, variables characterized by very different spatial and temporal characteristic scales. Using strongly coupled data assimilation also requires having access to much more computer power and developing less computationally expensive assimilation schemes. Work is progressing in both areas: weakly coupled data assimilation has been used to generate reanalyses with low resolution, and it is expected to be used in operational numerical weather prediction in the forthcoming years. Work is also progressing to test strongly coupled methods.

A second area that could lead to improvements is the move toward seamless coupled ensembles of analyses and forecasts, which would make the estimation of initial and forecast uncertainties more consistent.

The third area is to move toward higher resolution, which will allow for resolving better the smaller and faster scales, and their interaction with the slightly smaller scales that are already simulated by the models. All scales are relevant in weather prediction, and errors propagate from the smallest to the larger scales. If we consider the current operational models, we should not forget that even if they use resolutions of about 10 km, they resolve in a realistic way only scales that are about 4 times that resolution, that is, about 40 km. This means, for example, that frontal dynamics or extra-tropical intense storms are still poorly resolved. The adoption of higher resolution (say, a grid spacing of about 1 km) in global models can have a strong impact on the prediction of synoptic scale features in the medium range, and of low-frequency variability (e.g., the North Atlantic Oscillation or European blocking) in the monthly time scale.

Work is progressing to test and implement higher resolution ensembles both in modeling and assimilation, to make it possible to predict these events. We expect the current trend of a resolution increase of a factor of 2 every 4–5 years to continue: given that the state-of-the-art ensembles have today a resolution of about 20 km, we expect that by 2030 (2040) we should have ensembles covering the medium range (forecast days 1–15), and possibly the monthly and seasonal range, with a resolution of about 5 km (2 km).

244 **WEATHER PREDICTION**

10.2 What is an Earth digital twin?

The term "digital twin" has been used by the European Union to visualize their goal to develop a highly accurate digital model of the Earth to monitor and predict the interaction between natural phenomena and human activities.

A digital twin of the Earth atmosphere, or of the ocean, is a very accurate, high-resolution model of these Earth systems. Since developing digital twins of the atmosphere and the ocean involves investments in people and computer power, and the recoding of the Earth system models so that they can run more efficiently on the next generation of supercomputers (which might have tens of millions of computing units, instead of the few hundreds of thousands used today), the European Union has decided to coordinate work in these different areas under a strategy named "Destination Earth" (DestinE).

DestinE aims to unlock the potential of digital modeling of the Earth systems at a level that represents a real breakthrough in terms of accuracy and local detail. The initial focus will be on the possibility of improving the modeling of the effects of climate change and extreme weather events, their socioeconomic impact, and possible adaptation and mitigation strategies. Numerical weather prediction will also benefit from DestinE, since DestinE should lead to computationally more efficient Earth system models and coupled assimilation schemes.

Accordingly to the DestinE program (see their website), its users will be able to access and interact with vast amounts of Earth system and socioeconomic data in order to:

- Perform highly accurate, interactive and dynamic simulations of the Earth system, informed by rich observational datasets and more accurate analyses/reanalyses: for example, allowing a focus on thematic domains of societal relevance such as the regional impacts of climate change, natural hazards, marine ecosystems, or urban spaces.

- Improve prediction capabilities to maximize impact: for example, to protect biodiversity, manage water, renewable energy, and food resources, and to mitigate disaster risks in a changing world.
- Support European Union policymaking and implementation: for example, to assess the impact of existing environmental policies and legislative measures and support future evidence-based policymaking.
- Exploit the potential of distributed and high-performance computing and data handling at an extreme scale: for example, through an interactive platform that will host complex digital twins and comprehensive toolkits to develop and operate analytics-based models, with full access to vast amounts of diverse data.

Europe's industrial and technological capabilities will be reinforced through, for example, the simulation and observation of the entire Earth system and the use of artificial intelligence for data analytics and predictive modeling.

10.3 Will we be able to continue to improve the quality of weather forecasts?

Yes, there is still a lot of room for further improvements, both in the short-range prediction of the small and fast scales, and in the long forecast range. Although we know that the forecast skill horizon is limited, we do not think that we have reached it yet. We expect to be able to further extend it, since there are many areas of potential development in modeling (use of higher resolution and better models), observation (use of more and better data), and assimilation (move toward coupled assimilation methods).

The last 40 years have seen single medium-range forecasts gaining about 1 day of predictability each decade and ensemble-based, probabilistic forecasts in the medium range gaining about 1.5 days per decade. Longer range forecasts of large and slow scales have seen improvements by 1 week per decade (think, e.g., of the prediction of organized convection in the tropical regions and the Madden-Julian Oscillation), or even longer (think about seasonal prediction of the warming/cooling of the tropical Pacific linked to El Niño and La Niña phenomena). We expect to be able to continue to further improve the prediction of local weather events and difficult variables such as cloud cover, wind speed, and precipitation in the short range, and of large-scale weather events on the monthly and seasonal time scale.

If we consider the medium range, Figure 10.1 shows the evolution of the ECMWF day 5 forecast skill from 2010 to today, for large-scale phenomena, represented by the 500 hPa geopotential height, the mean sea-level pressure (MSLP), and the 850 hPa temperature, and for small-scale phenomena, represented by the 10 m wind and the total cloud cover. The

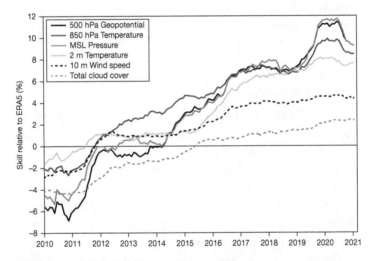

Figure 10.1. Evolution of skill of the HRES forecast at day 5, expressed as relative skill compared to ERA5. Verification is against analysis for 500 hPa geopotential, 850 hPa temperature, and mean sea-level pressure (MSLP), using error standard deviation as a metric. Verification is against analyses for the 500 hPa geopotential height, the 850 hPa temperature and the MSLP, and is against SYNOP for 2 m temperature, 10 m wind speed, and total cloud cover. (Source: ECMWF)

figure shows the relative skill of forecasts issued by the operational forecast system with respect to reforecasts generated using the ERA5 model starting from the ERA5 reanalyses. From this metric we can deduce the speed of forecast improvement for these different variables, and scales, since it compares the error of forecasts run with a fixed model, ERA5, against the operational forecasts generated with models and assimilation schemes that were updated about once every year.

ERA5 reanalyses and reforecasts have been generated using the same model and data assimilation: since ERA5 started being generated in 2014, the ERA5 model is very similar to the operational model used in 2014. Thus, we should not expect differences between the two operational and the ERA5 forecasts in 2014: indeed, there is no difference between the forecast skill for the vertically integrated variables 500 hPa geopotential height and mean sea-level pressure. There are some differences in other variables because although at that time the model was the same, the number of vertical levels was different, and this can have an impact on the skill of single-level variables, or on surface variables.

If we consider the 500 hPa geopotential height line (medium gray solid), Figure 10.1 shows that the relative skill (of the operational forecasts with respect to ERA5 reforecasts) was –6% in 2010 and climbed to +10% in 2021, thus indicating about a 1% relative-skill improvement every year. The difference in skill is smaller for the variables that are more affected by the small and fast scales, the 10 m wind speed (dark dashed line) and the total cloud cover (light dashed line), which improved by about 0.5%. Improvements have been slower for these variables because they are affected by the scales that are still not properly represented by global models with a resolution of about 10 km and an effective resolution of about 40 km. In the future, by improving the models and increasing resolution to about 2.5 km, we expect global models to be able to further improve the forecasts of these variables and add to the list shown in Figure 10.1 the ones that are not yet able to accurately predict at day 5 (e.g., precipitation type or aerosol concentration).

If we now consider the monthly forecast range, one area where we expect improvements is the prediction of synoptic patterns (spatial scales order 500–1,000 km, characteristic time of 3–5 days) over Europe. These patterns determine cold spells, and dry or wet periods during winter. Today, during regime transitions they can be predicted only about 7–10 days before they occur. Models still have difficulties in predicting the transitions from one type of circulation to another, for example, from zonal flow leading to wet conditions over central Europe, to high-pressure block conditions leading to dry and cold spells. Forecast errors could be linked to failures in the simulation of the interaction between the extra-tropical jet stream and the large-scale flow over Europe, with large-scale waves that propagated from the tropics, where they have been generated by organized convection. A possible second source of forecast error could be the failure to simulate correctly the impact of Greenland and of the Arctic sea ice on the polar circular and the jet stream. Improvements in the simulation of these aspects are expected to lead to an extension of the forecast skill horizon.

10.4 Will we ever be able to issue a "perfect" forecast?

No. There will always be initial condition and model uncertainties that will grow and affect the forecast quality.

Observations are characterized by an observation error, models represent real processes in an approximate way, and initial conditions are estimated applying techniques based on statistical assumptions: these are the main sources of numerical weather forecasts. We can aim to reduce the errors and improve our estimation of the forecast uncertainties, but we cannot eliminate them. We will further extend the forecast skill horizon for all scales, but uncertainty will always be part of our forecasts: we cannot eliminate uncertainty, but we can improve its estimate. We can aim to extend the forecast range up to which we can issue skillful and reliable ensemble-based probabilistic forecasts, but we will never be able to issue perfect forecasts. We can aim to tame the butterfly effect, but not to remove it!

10.5 In 2050, will we be able to predict the local weather of the next season?

In 2050, we will not be able to predict the local weather of the next season, but we expect to have improved substantially the prediction of large-scale patterns. We expect to continue to improve medium-range forecasts of the local weather by about 1–1.5 days every decade, which would extend the forecast skill horizon for fast, small spatial-scale phenomena (that characterize the local weather) from a few days to about 2 weeks, but not longer.

Today, ensembles can generate accurate and reliable forecasts of the local weather (e.g., surface wind, precipitation, and cloud cover at a specific location) a few days ahead and of large-scale features weeks (and for some features, months) ahead. The last decades have seen predictability gains of ensemble probabilistic forecasts of small and fast-scale phenomena of about 1.5 days per decade. This rate of improvement could continue, but it is difficult to see how it could be accelerated so much to give us accurate and reliable local weather forecasts one month or one season ahead.

Monthly and seasonal ensemble forecasts of large-scale phenomena could continue to be improved, for example, by adopting strongly coupled assimilation methods to reduce the uncertainties in the initial conditions of the slowly evolving Earth system components (ocean, sea ice, and deep soil), and by simulating in a better way the coupled processes in the forecast models. The last three decades have seen improvements in the prediction of large-scale phenomena on the monthly time scale (e.g., the Madden-Julian Oscillation in the tropics or the North Atlantic Oscillation patter in the extra-tropics) of about

3–5 days per decade, and on the seasonal time scale of about 1 month per decade.

We expect to be able to further extend the skill of the prediction of these phenomena, since we know that there are weaknesses in the way we simulate processes at the interface of Earth system components that could be improved. One example is the interaction between the stratosphere and the troposphere, another one is the interaction between land processes (e.g., orography and sea ice) and the low-level flow, and a further one is the interaction of thin cloud layers with radiation. The use of increased horizontal and vertical resolution, planned to reach 1–2 km in the operational ensembles in the next 20 years, and improvements in the simulated processes should deliver substantial predictability gains in this forecast range.

Thus, the forecast skill horizon for these large-scales will be further extended, but the monthly and seasonal forecasts of the local details (e.g., whether it will rain in 45 days from now at our location) will continue to be affected by large uncertainties.

10.6 Can artificial intelligence lead to improved predictions?

Yes, it could. Work is progressing to investigate how artificial intelligence could help in some specific areas: for example, it could lead to better and more efficient physical parameterization schemes and data assimilation schemes, and calibrated products.

With the term "artificial intelligence," we mean methods and software tools based on artificial neural networks and machine learning—in other words, models based on sophisticated statistical methods that rely on long training datasets to develop simulation models.

Although it is difficult to envisage that a forecast model based only on artificial intelligence could beat a numerical model deduced from the laws of physics, we know that parts of the physical parameterizations require very time-consuming calculations, and that artificial intelligence methods could help in developing surrogate schemes that are more efficient, and thus could allow the adoption of higher resolution.

Data assimilation is also based on sophisticated statistical models and requires inverting high-dimensional matrices. Artificial intelligence could lead to computer savings in these computations, and savings could make it possible to assimilate more observations: today we assimilate about 15% of the collected observations, and we should aim to increase this percentage as much as possible. Artificial intelligence could improve our observation operators, which are used to translate instrumental observations into model variables (e.g., they could help us use in a more effective way satellite images taken at different times to deduce atmospheric winds).

Another area where artificial intelligence could help in improving efficiency is in product generation, especially of calibrated ensemble-based products, that require the use of many forecast and reforecast data. Artificial intelligence can also lead to the development of new methods that can work with multi-variables at the same time.

10.7 What is an "environmental prediction model"?

An environmental prediction model is an Earth system model capable of simulating and predicting not only the canonical weather variable, but also chemical species and aerosols that can be used to monitor and predict environmental conditions.

The inclusion in Earth system models of schemes that simulate the few chemical reactions that govern the concentration of key pollutants and greenhouse gases has made it possible for the Earth system models to compute their concentrations and predict their evolution. For example, it has allowed for monitoring and predicting the concentration of pollutants, such as carbon monoxide (CO), nitrogen dioxide (NO_2), ozone (O_3), particulate matter (PM10, PM2.5), and sulfur dioxide (SO_2).

The inclusion of a carbon cycle could also allow for monitoring and predicting the concentration of greenhouse gases, such as carbon dioxide (CO_2) and methane (CH_4).

The European Union Copernicus Atmospheric Monitoring Service (CAMS) is an example of such a global model and data assimilation system, constructed by coupling the ECMWF Earth system model with complex chemical models developed by other research institutes. The result has been a data assimilation and prediction system capable of monitoring and predicting, in addition to weather variables, aerosols, greenhouse gases, and ozone. Since the chemical models are computationally very heavy, the resolution used by the CAMS operational system is lower than the one used by ECMWF for its operational weather prediction, about 25 km instead of 10 km. Furthermore, CAMS issues forecasts valid only for 5 days.

CAMS has also produced an environmental global reanalysis that covers the past 19 years, from 2003 to date, that can be used to track the trend in chemical species and particulate.

As has been the case for weather prediction, the development of environmental systems has been possible, thanks to major investments in new observing systems that have allowed for the collection of good-quality observations of chemical species and aerosols. About 80% of the European Union Copernicus funding has been invested to develop and launch a set of satellites, the "Sentinels": Sentinel-1A was launched in 2014, followed by Sentinel-1B (2016), 2A (2015) and 2B (2017), 3A (2016), and 3B (2018). Sentinels 1, 2, and 3 probes are in polar orbits and monitor a variety of Earth processes, including sea-surface height, sea-ice temperatures, ocean surface wind speeds, atmospheric aerosols, vegetation on land, and snow and ice coverage on land. Sentinels 4, 5, and 6 are still to be launched: 4 and 5 will monitor atmospheric composition, and 6 will use a radar altimeter to monitor the global sea-surface height alterations related to climate change with unprecedented accuracy.

Future upgrades and resolution increases of the CAMS model and data assimilation could allow for the identification of key polluters and emitters of greenhouse gases in real time. In fact, the model used to predict the forward evolution in time of initial concentrations of chemical species (including greenhouse gases like CO_2 and CH_4) can also be used backward in time, to identify the source of emissions of enhanced concentration of chemical species observed by satellite observing platforms. This will lead to the generation of new types of environmental products that will help us make a transition toward a future with less and zero greenhouse gas emissions, and less pollution.

10.8 Is weather prediction evolving into environmental prediction?

Yes. Actually, weather prediction has been evolving in that direction already for a few years, as the demand for a new type of data that allow for monitoring and predicting the state of the environment keeps increasing.

Forty years of advances in weather prediction have made it possible to issue good-quality predictions of the physical variables that drive chemical reactions (temperature, radiation, air density, and composition). This has led to the inclusion, in weather prediction models, of modules that simulate chemical reactions, and this has allowed for simulating in a realistic way the evolution of chemical species. Environmental models, such as the European Union Copernicus Atmospheric Monitoring Service (CAMS), produce good-quality forecasts of some chemical species and aerosols up to forecast day 5.

In the past 4 years, the European Union CO_2 Human Emissions (CHE) project has been coordinating efforts toward the development of a monitoring capacity for anthropogenic CO_2 emissions. This challenging target has been aligned with the European Commission's objective to enhance modeling capabilities for CO_2 fossil fuel emissions, along with other natural and anthropogenic CO_2 emissions and transport. The CHE project has been integrating the monitoring system components, as well as innovation for estimating fossil fuel CO_2 fluxes. These include reconciling bottom-up and top-down constraints and handling systematic errors of satellite sensors. Earth observations from satellites have been combined with in situ CO_2 observations and information from co-emitters or isotopes to support the attribution of fossil fuel emissions and uncertainty reduction. CHE will ensure the transfer of science and technology requirements and recommendations for strengthening existing assets with a view to developing an anthropogenic CO_2 monitoring service.

The future will see operational environmental predictions such as CAMS benefiting from projects like CHE and further expanding the range of chemical species that they can simulate realistically. This should lead to an extension of the forecast range covered by CAMS ensembles.

10.9 As global models keep increasing resolution, will we still use limited-area models?

Yes, the demand for very fine and detailed information in the short forecast range (say, up to 3 days) will still be present in the future, and the only way to satisfy this demand will be to run higher-resolution, limited-area models nested into global models.

Good-quality limited-area models, run with a finer resolution than the global models into which they are nested (say, with a grid spacing about 5–10 times finer than global models), have proven to provide more accurate forecasts of the very fine and fast scales than the global models into which they have been nested. This has been the case especially if their initial conditions have been generated using also a higher-resolution assimilation system, capable of assimilating local, higher-resolution observations not included in global assimilation systems. In particular, the localization and intensity of extreme weather events, such as wind storms and precipitation events in areas with strong orographic forcing, or the prediction of the correct timing of tropical convection on islands has been shown to be more accurate in limited-area forecasts than in global forecasts.

Today, the top-quality, state-of-the-art global and limited-area models have resolutions of about 10 and 2 km. Some national meteorological centers follow already a double nesting approach, where two limited-area models with increasing resolution are nested in the global model, to provide a forecast chain with resolutions of 10, 2, and about 0.5 km.

The move toward the use of more alternative sources of electricity based on solar energy and wind has made the electricity production more weather dependent. Electricity demand is also weather dependent: it is linked to temperature, rainfall, and cloud cover. This has led to an increased demand for more accurate and detailed weather forecasts, which could be used to predict how much electricity could be produced in the next hours and few days. Limited-area models, thanks to their higher resolution compared to global models, are capable of predicting the local details of solar irradiance (that depends from the cloud cover) and wind speed more accurately then the global models.

We expected that when, in the near future, global models increase resolution to 5 and then 1 km, limited-area models will increase their resolution to 1 and then 0.2 km. Until nested limited-area models continue to provide better local forecasts of the small and fast scales, we expect to continue to see them used in operations, with resolutions of a factor of about 5–10 finer than the global models into which they are nested.

10.10 Would a future operational suite look very different from today's?

Not really. We will see a move toward the use of ensembles of analyses and forecasts generated using Earth system models, with analyses and short-range (up to 3-day) forecasts generated more often than today (say, in a quasi-continuous way, up to every hour), medium-range (3- to 15-day) forecasts issued twice a day, monthly extensions issued once a day, and seasonal extensions once a week.

We do not expect the key processes of numerical weather prediction (observation collection and quality control, data assimilation, numerical integration, and forecast products generation) to change dramatically in the forthcoming decades, but we do expect improvements in forecast skill on all spatial and temporal scales, thanks to increased resolution, better simulation of relevant processes, and the use in operation of coupled data assimilation schemes. Artificial intelligence might substitute parts of the parameterization codes of the models, and it could lead to advances also in assimilation and product generation.

We expect that all the leading prediction centers will have a suite of seamless ensembles that will be used to generate probabilistic analyses and forecasts. Ensemble of analyses and short-range forecasts could be produced in a more continuous way, say every 3 hours or possibly even every hour, to exploit the continuous influx of observations in a more timely manner. The use of much higher resolution in the monthly and seasonal time scale is expected to lead to major advances in the prediction of large-scale patterns, especially over Europe, which remains a very difficult region for this time scale, and during transitions between different regimes.

Earth system model versions are expected to include upgrades in deep soil and land surface schemes (e.g., by simulating in a better way the impact of cities on the atmospheric flow and dynamic vegetation), aerosols, and at least a few, key chemical species, so that they can provide environmental forecasts. Reforecasts will routinely be produced and used to calibrate forecast products. The use of the same model version across scales and forecast ranges will allow more thorough diagnostics, and this could lead to model improvements. Product generation will also benefit from new approaches, for example, based on machine learning, to generate new, calibrated products. More environmental variables will be predicted, as part of the operational suites.

Thus, although the process followed to generate forecasts will be the same (observation collection and quality control, data assimilation, numerical model integration, product generation), it will be executed in a more efficient way, in a more timely manner, using more realistic and reliable ensembles of Earth system models and assimilations.

10.11 Key points discussed in Chapter 10 "A look into the future"

These are the key points discussed in this chapter:

- A further move toward Earth system approaches to modeling, the adoption of more seamless approaches in data assimilation and prediction, and increased resolution are three key areas of development of weather prediction.
- Digital twins of the atmosphere and ocean components of the Earth systems, that is, higher resolution and more accurate replica of the real components, are being developed, to further improve weather and climate prediction.
- Although the forecast skill limit is finite, there is still room for further improvements, both in the short forecast range, in the prediction of the fast and small spatial scales, and in the long forecast range, in the prediction of low-frequency phenomena.
- Forecasts will never be perfect and will always be uncertain: we can improve their accuracy and reliability, and expand the forecast skill horizon, but perfection is impossible.
- We should be able to continue to expand the forecast skill horizon in the next two decades, at a rate of about 1–1.5 days per decade in the medium range (lead times of up to 2 weeks), 5–7 days per decade in the monthly

time range, and 1 month per decade in the seasonal time range.

- As weather forecasts keep improving, environmental prediction models, capable of monitoring and predicting the evolution of chemical species and aerosols, will be developed and used in operational daily production.
- It is expected that the trend from weather-only to environmental prediction will continue, as the demand for an expansion of the forecast information continues.
- Higher-resolution limited-area models, nested in lower-resolution global models, are expected to continue to be part of the operational suite of the leading operational meteorological centers.

ESSENTIAL GLOSSARY

Analysis State of the atmosphere at a specific time, computed using data assimilation procedures by merging observations and a first-guess estimate of the state of the atmosphere, usually defined by a short-range forecast. The analysis defines the initial conditions from where a numerical weather prediction is computed.

Anomaly (of a field) The difference between the value of the field at a specific time and its time average over a long time period; for example, when we say that in 2020 global warming was about 1.2°C, we mean that the 2020 mean surface temperature anomaly with respect to the 1850–1900 average is 1.2°C (more precisely, 1.2°C is the difference between the annual mean temperature in 2020 and the mean of the 1850–1900 period).

Anomaly correlation coefficient (ACC) A measure of forecast accuracy: it depends from how close the predicted anomaly is to the observed anomaly. A perfect forecast has an ACC of 1.

Anticyclone and cyclone circulations These are two types of large atmospheric circulation patterns: anticyclonic circulations, identified on weather maps with the letter "H" (for high pressure) and characterized by a clockwise atmospheric flow, and cyclonic circulations, identified on weather maps by the letter "L" (for low pressure) and characterized by an anticlockwise atmospheric flow.

Arctic oscillation (AO) A large-scale, hemispheric-wide circulation around the Arctic, characterized by a back-and-forth shifting of atmospheric pressure between the Arctic and the mid-latitudes of the North Pacific and North Atlantic. When the AO is strongly positive, a strong mid-latitude jet stream steers storms northward, reducing cold air outbreaks in the mid-latitudes.

Butterfly effect In chaos theory, it indicates a sensitive dependency on initial conditions: in weather, it means that even the flap of a butterfly wing can affect the global weather state. Its definition is closely associated with the work of Edward Lorenz.

Bytes, megabytes, and terabytes The byte is a unit of digital information that most commonly consists of 8 bits (but it can include between 1 and 48 bits). Historically, the byte was the number of bits used to encode a single character of text on a computer, and thus it is the smallest addressable unit of memory in many computer architectures. A megabyte is one million (10^6) bytes, and a terabyte is one trillion (10^{12}) bytes.

Central processing unit (CPU) The portion of a computer that retrieves and executes instructions.

Chaotic behavior A system shows chaotic behavior if two orbits of the system that are very close at a certain time will diverge at a certain time in the future, independently of how close they are initially. A system with a chaotic behavior is a system with orbits that are very sensitive to initial conditions.

Climate It describes the statistics of the weather phenomena computed over a long period, say at least a month or a season, but more correctly a few years.

Cluster It is an ensemble of objects, grouped together because they are similar. In ensemble weather prediction, a cluster forecast is defined by the ensemble members that are close to each other and are further away from the rest of the members not included in the cluster. A cluster is usually represented by a "representative member"—that is, one of its members that is closest to the members of the cluster and further away from the members of the other clusters. Clusters are a way to condense the information contained in ensembles.

Coordinated Universal Time (UTC) The primary time standard of the world, within 1 second of the mean solar time at 0° longitude. It is not adjusted for daylight savings time.

Coriolis force In physics, it is an inertial, of fictitious force, that acts on objects in motion within a rotating frame of reference, as it is the case in Earth system modeling applied to weather and climate.

Coupled model In weather and climate science, it indicates a model that simulates not just one component of the Earth system (e.g., the atmosphere or the ocean), but that includes different components (e.g., the atmosphere, the ocean, the land, and the cryosphere).

Cut-off low-pressure system A small-scale, cyclonic circulation area with a deep low central pressure that appears on meteorological maps as one or more concentric circles.

Data assimilation A procedure followed in meteorology to merge in an optimal way observations collected over a time window (usually spanning a few hours) with a first-guess state of the atmosphere valid for the same time window, so that the merged state is closer to the observations than the initial first guess. The first guess is usually defined by the most recently available forecast.

Degrees Celsius and Kelvin The Kelvin is the primary unit of temperature in the International System of units (SI), and the absolute scale of temperature is expressed in degrees Kelvin (K). X degrees K correspond to $(X - 273.15)$ °C; zero degree K is the absolute zero temperature, and it corresponds to −273.15°C.

Degrees of freedom (DOF) (of a system) Identifies how many independent variables are required to fully describe the state of the system.

Dynamical model A mathematical model of a system that describes its time evolution.

Earth system model A mathematical model of the Earth that includes the key processes of the atmosphere, ocean, cryosphere, and land components.

El Niño and La Niña events Large spatial-scale coupled ocean-atmosphere phenomena that occur in the tropical Pacific: El Niño (La Niña) is characterized by warmer (colder) than average sea surface temperature in the central and eastern central Pacific. Due to their large spatial scale and long characteristic time (few months), they affect the global circular patterns and probabilities of extreme events (e.g., hurricanes in the Atlantic and Pacific tropical ocean basins).

Ensemble, or ensemble forecast system An ensemble, or an ensemble forecast system, includes many single forecasts, defined so that they sample the initial uncertainties and simulate all relevant sources of forecast uncertainty.

First guess An estimate of the state of the atmosphere used in data assimilation, usually defined by the most recently available forecast.

Flops, teraflops, and exaflops Flops, the number of floating point operations per second, are a measure of computer performance. Teraflops are 10^{12} flops, petaflops are 10^{15} flops, and exaflops are 10^{18} flops. The supercomputers used at the time of writing (2022) in meteorology have a sustained performance of between 5 and 50 petaflops.

Forecast error and forecast skill The forecast error is a measure of the distanced between a forecast and reality, while the forecast skill is a relative measure of forecast error with respect to a reference forecast (usually defined by persistency or climatology). A forecast is said to be skillful if it has an error that is smaller than the reference forecast.

Forecast ranges They are usually grouped in the short range (1 to 3 days), medium range (3 to 15 days), extended or monthly or subseasonal range (2 to 8 weeks), and long or seasonal range (1 to 12 months).

Greenhouse gases (GHGs) Chemical species contained in the atmosphere that absorb long-wave radiation and thus heat up the atmosphere. Apart from water vapor, the three main greenhouse gases are carbon dioxide (CO_2), methane (CH_4), and nitrous oxide (N_2O).

Greenwich Mean Time (GMT) The time displayed by the Shepherd Gate Clock at the Royal Observatory in Greenwich, London.

High- and low-frequency phenomena A high-frequency (low-frequency) phenomenon has a very fast (long) temporal characteristic time. For example, convection is a high-frequency phenomenon, while El Niño is a low-frequency phenomenon.

High- and low-level pressure systems Two types of large atmospheric circulation patterns: a high-pressure system (identified on weather maps by "H") is characterized by a clockwise atmospheric flow, while a low-pressure system (identified on weather maps by the letter "L") is characterized by an anticlockwise atmospheric flow.

Intergovernmental Panel on Climate Change (IPCC) The United Nations body created in 1998 by the World Meteorological Organization (WMO) and the United Nations Environment Program (UNEP) to assess the science related to climate change. It prepares comprehensive Assessment Reports about the state of scientific, technical, and socioeconomic knowledge on climate change, its impacts and future risks, and options for reducing the rate at which climate change is taking place.

Laws of thermodynamics The laws of thermodynamics define a set of physical quantities (temperature, energy, entropy) that characterize the thermodynamic state of a system. The zeroth law of thermodynamics defines thermal equilibrium. The first law states that, when energy passes in or out of a system, the system's internal energy changes in accordance to the law of conservation of energy. The second law states that in a thermodynamic process, the sum of the entropies of the interacting system never decreases. The third law states that a system's entropy approaches a constant value as the temperature approaches the absolute zero.

Limited area model (LAM) A model that describes the time evolution of the Earth system inside a finite domain (as compared to global models that cover the whole globe)

Madden-Julian Oscillation (MJO) A large-scale (a few thousand kilometers) organized convection phenomenon that occurs in the tropics. It has a characteristic time of 2–3 weeks, and due to its large scale, it can affect the global circulation.

Moist processes Physical processes that involve water (H_2O) molecules.

Newton laws of motion They are the three basic laws of classical mechanics that describe the relationship between the motion of an object and the forces acting upon it. The first law states that a body remains at rest or in motion with a constant speed in a straight line, unless acted upon by a force. The second law states that when a body is acted upon by a force, the time rate of change of its momentum (the product of its mass time velocity) equals the force. The third law states that if two bodies exert forces on each other, these forces have the same magnitude but opposite directions. They were stated in Newtown's *Philosophiae Naturalis Principia Mathematica*.

North Atlantic Oscillation (NAO) A large spatial-scale atmospheric pressure seesaw in the North Atlantic region. The common pressure features seen in the North Atlantic Ocean are for large regions of relatively high pressure centered over the Azores islands (west of Portugal, known as the subtropical or Azores high) and low pressure centered over Iceland (the subpolar or Icelandic low). The NAO describes the relative changes in pressure between these two regions (Azores minus Iceland) and was discovered in the 1920s by Sir Gilbert Walker. The NAO has a strong influence on winter weather and climate patterns in Europe and North America, and it leads to changes in the intensity and location of the North Atlantic jet stream

Omega blocked pattern A large-scale circulation pattern characterized by a high-pressure system, with two low-pressure systems to the west and the east, causing the atmospheric flow to depict an omega shape. It has a rather long characteristic time (a few days) and can affect the large-scale circulation over Europe for many days.

Orbits (of a system) When we study dynamical systems, an orbit is defined by the collection of points related to the system time evolution. An orbit is computed by integrating in time the equations that describe the system dynamics from an initial state, and on maps it is shown by a continuous line.

Pacific North America (PNA) pattern A large spatial-scale pattern that characterizes the circulation over the North Pacific Ocean and the North American continent with two modes, a positive and a negative phase. The PNA is associated with strong fluctuations in the strength and location of the East Asian jet stream.

Potential economic value (PEV) of a forecast A measure used to assess the skill of a forecast based on a simple cost/loss model, defined as a potential saving that it can lead to if used to take weather-related decisions.

Probability density function (PDF) A function that describes the probability of possible different weather events.

Ranked probability score (RPS) and continuous ranked probability score (CRPS) Most commonly used metrics to assess the quality of a probabilistic forecast. They are the equivalent of the root mean square error for single forecasts.

Reanalyses and reforecasts A reanalysis is an analysis of a past state of the atmosphere recomputed with today's data assimilation and model systems. A reforecast is a forecast of a past case recomputed using today's model starting from a reanalysis. Reanalyses are extremely valuable datasets that cover many decades: since they are computed using the same model and data assimilation (while operational analyses are computed with systems that are routinely upgraded, say about once a year), they provide a consistent state dataset that can be used to assess the evolution of the Earth system. Reforecasts also provide very valuable datasets of forecasts generated using the same system but covering cases of different years (while operational forecasts are generated with systems that are routinely upgraded).

Reliable ensemble Reliability is a property of an ensemble forecast system. An ensemble is reliable if, over a large number of cases, when it predicts that an event has a probability p to occur, the event occurs with a frequency $f = p$. Reliability can be assessed using a reliability diagram that contrasts the forecast probabilities and the frequencies of occurrence: a reliable ensemble has all its points (p,f) lying on the diagonal of such a scatter diagram. Reliability can also be measured by contrasting the average (over many cases) ensemble spread and the average error of the ensemble-mean forecasts: the two average values should superimpose.

Root mean square error (RMSE) It is the most commonly used metric for computing the forecast error: it is defined as the root mean square of the average of the squared difference between the forecasts and the observations.

Rossby deformation radius It is the length scale at which rotational effects (due to the rotation of the Earth) become as important as buoyancy or gravity wave effects in the evolution of the flow about some disturbance. In the atmosphere, the Rossby deformation radius is about 1,000 km, which is similar to the scale of synoptic features (cyclonic and anticyclonic patterns). In the ocean, it varies between 200 km at the equator and 10 km at high latitudes: eddies in the ocean vary similarly, with typical scales of about 200 km at the equator and about 10 km or less at northern latitudes.

Satellite active and passive sensors Satellite instruments can be characterized as active and passive. Active sensors provide their own source of energy to illuminate the object they observe; the sensors emit radiation at a specific wavelength toward the object and then observe the radiation that is reflected or backscattered. Passive sensors detect the natural energy that is emitted, or reflected, by the object they observe; reflective sunlight is the most common source of radiation measured by passive sensors.

Skill and skill score The skill of a forecast measures how accurate the forecast is compared to a reference. The skill score is a metric used to measure the skill: given a score SC that measures the error of a forecast, the skill is defined as the ratio between the difference of the score of the forecast minus the score of a reference [SC(f) – SC(ref)], divided by the difference between the score of a perfect forecast minus the score of the reference [SC(perf) – SC(ref)].

Stable equitable error in probability space score (SEEPS) A negatively oriented score with values between 0 and a maximum expected value of 1 (over a sufficiently long period) for an unskilled forecast.

State variable A variable used to describe the characteristics of a system. For the atmosphere, state variables are temperature, surface pressure, wind, density, and humidity.

Stefan-Boltzmann law of physics Describes the power radiated from a black body in terms of its temperature.

Sustained performance The amount of useful work a computer can produce in a given amount of time on a regular basis. It is one of the parameters used to characterize computer power.

Synoptic scale systems Atmospheric weather systems with a characteristic spatial scale of about 1,000 km and a characteristic time scale of a few days. Most of the high- and low-pressure systems are synoptic scale systems.

Three-dimensional grid Mesh used to solve numerically the equations that describe the motions of the atmosphere. It is defined by its horizontal grid spacing (between 10 and 50 km for global model used in operational weather prediction in 2022), the position of the top of the atmosphere (between 50 and 80 km for global weather models), and the number of vertical levels (between 50 and 140 for global weather models).

Time step The minimum difference between two consecutive times at which the equation of motions of the Earth system are computed. It depends on the resolution of the model and the numerical schemes used to solve the equations numerically.

Trough system It is an elongated area of relatively low pressure extending from the center of a region of low pressure. Air in a high-pressure area compresses and warms as it descends. This warming inhibits the formation of clouds, meaning the sky is normally sunny in high-pressure areas.

FURTHER READING

Listed here are a few recent books in which the interested reader can find more detailed information on the topics discussed in this book.

Andina, T., and Corvino, F. 2023. *Global Climate Justice: Theory and practice.* E-International Relations Publishing.

Daley, R. 1993. *Atmospheric Data Assimilation.* Cambridge University Press.

Duan, Q., Pappenberger, F., Thielen, J., Wood, A., Cloke, H., and Schaake, J. 2018. *Handbook of Hydrometeorological Ensemble Forecasting.* Springer.

Hartman, D. L. 2016. *Global Physical Climatology.* 2nd ed. Elsevier.

Holton, J. R., and Hakim, G. J. 2012. *An Introduction to Dynamic Meteorology.* Elsevier.

Hoskins, B. J., and James, I. N. 2014. *Fluid Dynamics of the Mid–Latitude Atmosphere (Advancing Weather and Climate Science).* Wiley-Blackwell.

IPCC. 2021. "Summary for Policymakers." In *Climate Change 2021: The Physical Science Basis. Contribution of Working Group I to the Sixth Assessment Report of the Intergovernmental Panel on Climate Change,* edited by V. Masson-Delmotte, P. Zhai, A. Pirani, S. L. Connors, C. Péan, S. Berger, N. Caud, Y. Chen, L. Goldfarb, M. I. Gomis, M. Huang, K. Leitzell, E. Lonnoy, J. B. R. Matthews, T. K. Maycock, T. Waterfield, O. Yelekçi, R. Yu, and B. Zhou, 31. Cambridge University Press.

IPCC. 2022a. *Climate Change 2022: Impacts, Adaptation, and Vulnerability.*
Contribution of Working Group II to the Sixth Assessment Report
of the Intergovernmental Panel on Climate Change. Edited by H.-O.
Pörtner, D. C. Roberts, M. Tignor, E. S. Poloczanska, K. Mintenbeck,
A. Alegría, M. Craig, S. Langsdorf, S. Löschke, V. Möller, A. Okem,
and B. Rama. Cambridge University Press.

IPCC. 2022b. *Climate Change 2022: Mitigation of Climate Change.*
Contribution of Working Group III to the Sixth Assessment Report
of the Intergovernmental Panel on Climate Change. Edited by P.
R. Shukla, J. Skea, R. Slade, A. Al Khourdajie, R. van Diemen, D.
McCollum, M. Pathak, S. Some, P. Vyas, R. Fradera, M. Belkacemi,
A. Hasija, G. Lisboa, S. Luz, and J. Malley. Cambridge University
Press. doi:10.1017/9781009157926.

James, I. 2010. *Introduction to Circulating Atmospheres.* Cambridge
University Press.

Kalnay, E. 2002. *Atmospheric Modelling, Data Assimilation and
Predictability.* Cambridge University Press.

Lackmann, G. 2012. *Midlatitude Synoptic Meteorology.* University of
Chicago Press.

Lorenz, E. 2004. *The Essence of Chaos.* University of Washington Press.

Olafsson, H., and Bao, J.-W. 2020. *Uncertainties in Numerical Weather
Prediction.* Elsevier.

Palmer, T. N., and Hagedorn, R., 2006. *Predictability of Weather and
Climate.* Cambridge University Press.

Palmer, T. N. 2023. The Primacy of Doubt: From climate change to
quantum physics, how the science of uncertainty can help predict
and understand our chaotic world. Oxford University Press, pp. 279
(ISBN 978–0–19–2843593).

Philander, S. G. 1989. *El Nino, La Nina, and the Southern Oscillation.*
Academic Press.

Robertson, A. W., and Vitart, F. 2018. *The Gap Between Weather and
Climate Forecasting: Sub-seasonal to Seasonal Prediction.* Elsevier.

Vallis, G. K. 2017. *Atmospheric and Oceanic Fluid Dynamics. Fundamentals
and Large-Scale Circulation.* Cambridge University Press.

Wallace, J. M, and Hobbs, P. V. 2006. *Atmospheric Science: An Introductory
Survey.* 2nd ed. Vol. 92 in the International Geophysical Series.
Academic Press.

Wilks, D., Vannitsen, S., Messner, J. 2018. *Statistical Post-processing of
Ensemble Forecasts.* Elsevier.

USEFUL LINKS

Following are links to the institutions or projects mentioned in this book.

- Centro Euro-Mediterraneo sui Cambiamenti Climatici: https://www.cmcc.it
- Copernicus project: https://www.copernicus.eu/en
- Copernicus Climate Change Service (C3S): https://climate.copernicus.eu
- Copernicus Atmosphere Monitoring Service (CAMS): https://atmosphere.copernicus.eu
- CO_2 Human Emissions (CHE) project, European Union: https://www.che-project.eu
- Destination Earth (DestinE) European Union project: https://www.destineproject.eu
- European Centre for Medium-range Weather Forecasts (ECMWF): https://www.ecmwf.int
- European Organisation for the Exploitation of Meteorological Satellites (EUMETSAT): https://www.eumetsat.int
- European Space Agency (ESA): https://www.esa.int
- EC Earth Consortium: https://ec-earth.org
- Global Monitoring Laboratory (GML) of the Earth System Research Laboratories (ESRL), Mauna Loa Observatory: https://gml.noaa.gov/obop/mlo/
- Grantham Institute for Climate Change and the Environment of Imperial College: https://www.imperial.ac.uk/grantham/
- Intergovernmental Panel on Climate Change (IPCC): https://www.ipcc.ch

- Istituto Studi Superiori IUSS of Pavia: https://www.iusspavia.it/en
- Leading meteorological centers issuing daily operational global forecasts:
  - National Oceanic and Atmospheric Administration (NOAA), United States: https://www.noaa.gov
  - China Meteorological Administration (CMA): http://www.cma.gov.cn/en2014/
  - Korea Meteorological Administration (KMA): https://www.kma.go.kr/eng/index.jsp
  - Meteo France: https://meteofrance.com
  - UK Meteorological Office: https://www.metoffice.gov.uk
  - Japan Meteorological Agency (JMA): https://www.jma.go.jp/jma/indexe.html
  - Centro de Previsao de Tempo e Estudos Climaticos (CPTEC), Brazil: https://www.cptec.inpe.br
  - Environment and Climate Change Canada (ECCC): https://www.canada.ca/en/environment-climate-change.html
  - Global Modelling and Data Assimilation (GMAO) of the National Aeronautics and Space Administration (NASA), United States: https://gmao.gsfc.nasa.gov
- National Snow and Ice Data Center (NSIDC), United States: http://nsidc.org/
- Our World in Data: https://ourworldindata.org
- Scuola Superiore Sant'Anna Pisa: https://www.santannapisa.it/en
- Scuola Normale Superiore of Pisa: https://www.sns.it/en
- Top-500 computer list: https://www.top500.org
- University of Reading, Department of Meteorology: https://www.reading.ac.uk/meteorology/
- World Bank Climate Knowledge portal: https://climateknowledgeportal.worldbank.org/download-data
- World Meteorological Organization (WMO): https://public.wmo.int/en

ABOUT THE AUTHOR

Roberto Buizza was born in Lecco, Italy, in 1961. He has a "Laurea" (master's degree) in physics from the University of Milano, with a thesis on plasma fusion; a PhD in mathematics from University College London, with a thesis on perturbation growth in the atmosphere; and a master's in business administration from London Business School, with a management report on Monte Carlo–based risk assessment.

Between 1987 and 1991, he worked at the "Centro di Ricerca Termica e Nucleare" of the Electricity Board of Italy (CRTN/ENEL), where he developed models to investigate and predict the diffusion of pollutants emitted by power stations, including from nuclear ones. At CRTN/ENEL he started working in meteorology, weather prediction, and climate science.

In 1991, he joined the European Centre for Medium-Range Weather Forecasts (ECMWF), where he had been a key developer of the ECMWF models. At ECMWF, he served in different capacities in the Research Department, including as Head of the Ensemble Prediction Section, Head of the Predictability Division, and Lead Scientist.

In November 2018, he joined "Scuola Superiore Sant'Anna" of Pisa as a Full Professor in Physics. Here, he established a new initiative on climate change and sustainability with the support of three of the six Italian Scuole Superiori, top-quality Italian research universities: Scuola Superiore Sant'Anna of Pisa, Scuola Normale Superiore of Pisa, and Istituto Studi Superiori IUSS of Pavia. The main outcome of this initiative has been a new national doctoral school on sustainability and climate change, which started in November 2021 with over 100 PhD students partially funded by 30 Italian universities and partially directly by the ministry of research.

In February 2022, he joined the Italian Embassy in London, as Scientific Attaché, where he has been working in science diplomacy.

Expert in numerical weather prediction, ensemble methods, and predictability, he has written approximately 250 scientific and technical publications, of which 117 are in the peer-reviewed literature, and contributed to 16 scientific books.

Since joining Scuola Superiore Sant'Anna, he has been very active in communicating climate science to the public and in promoting initiatives aiming to achieve immediate and impactful actions to reduce greenhouse gases' emissions and deal with climate change. On these topics, he has written articles in Italian national newspapers and has been interviewed by the main Italian national television channels.

He has a wonderful family, with a fantastic wife, three great grown-up children, and two very affectionate dogs. (I should also mention the cat, who used to inspire me when I started writing this book, and who died just before it was finalized.)

INDEX

For the benefit of digital users, indexed terms that span two pages (e.g., 52–53) may, on occasion, appear on only one of those pages.

Figures are indicated by *f* following the page number